Michael Betz

The CERN Resonant WISP Search (CROWS)

Karlsruher Forschungsberichte aus dem
Institut für Hochleistungsimpuls- und Mikrowellentechnik

Herausgeber: Prof. Dr.-Ing. John Jelonnek

Band 5

The CERN Resonant WISP Search (CROWS)

by

Michael Betz

Dissertation, Karlsruher Institut für Technologie (KIT)
Fakultät für Elektrotechnik und Informationstechnik, 2013
Referent: Prof. Dr. rer. nat. Dr. h.c. Manfred Thumm
Korreferent: Prof. Dr. Oliver Keith Baker

Impressum

 Scientific
Publishing

Karlsruher Institut für Technologie (KIT)
KIT Scientific Publishing
Straße am Forum 2
D-76131 Karlsruhe

KIT Scientific Publishing is a registered trademark of Karlsruhe
Institute of Technology. Reprint using the book cover is not allowed.

www.ksp.kit.edu

Print on Demand 2014

ISSN 2192-2764
ISBN 978-3-7315-0199-2
DOI: 10.5445/KSP/1000039767

Foreword of the Editor

The "CERN Resonant WISP Search" (CROWS) experiment looks for the existence of Weakly Interacting Sub-eV Particles (WISPs) using microwave techniques. Subject of this PhD thesis is the design, implementation and first measurements of the CROWS experiment.

Axion Like Particles (ALPs) and Hidden Sector Photons (HSPs) are two members of the WISP family. Their existence could reveal the composition of cold dark matter in the universe and explain a large number of astrophysical phenomena. Despite their strong theoretical motivation, the hypothetical particles have not been observed in any experiment up to now. One way to probe the existence of WISPs is to study their interaction with photons in a so-called "light shining through the wall (LSW)" experiment. A laser beam is guided through a strong magnetic field ("emitting region") and provides photons, which can convert into hypothetical ALPs or other WISPs. HSPs can even be generated without a magnetic field. The photons are blocked by a wall but the WISPs practically do not interact with matter and, hence, can propagate to the "detection region". In the detection region they can reconvert into photons, which subsequently can be detected. A hypothetical WISP would reveal itself as light, apparently shining through the opaque wall.

The CROWS experiment of Dr.-Ing. Michael Betz is based on the same principles as the LSW experiment, but operates in the microwave regime at a frequency of several GHz, replacing the laser by a signal generator and the optical detector by a sensitive microwave receiver. CROWS uses the fact, that microwave cavity resonators are technologically easier to implement than optical Fabry-Pérot ones. On the other hand, proper isolation of the EM fields in each of the two cavities is one of the primary concerns to achieve the sufficient detection sensitivity. It requires an isolation in the order of 300 dB to stay below the detection threshold. The shielding concept considers several, demountable shells, providing about 100 dB of attenuation each. That allows characterization of each shell by conventional measurement techniques. Additionally, a test signal is emitted into the shells which ensures that the isolation performance is not deteriorated during each experimental run e.g., due to vibrations or corrosion. Two shielding enclosures have been built, one for the detecting cavity, which for the ALP experiments was placed within a 3 T magnet and one for the components of the signal receiver, which was placed outside the magnet. Transmission of analog signals between the two shielding enclosures has been accomplished by optical transceivers. A homemade double heterodyne signal receiver and a commercial signal analyzer have been utilized to record a 20 kHz or 2 kHz wide slice of the signal spectrum, respectively, at which a WISP signal would have been expected. The power spectrum of the recorded time-trace has been evaluated, which corresponds to narrowband filtering. For a measurement run of 29 hours, the resolution bandwidth of <10 µHz has been achieved. This allows detection of signals with less than -210 dBm. Taking a 3 GHz signal, that corresponds to a flux of around 0.5 photons/s. Both receivers were optimized for minimum frequency drift by phase locking of all oscillators to a common reference clock.

HSPs or ALPs were not observed in the most sensitive measurement-runs of the CROWS experiment. For HSPs with a mass corresponding to 10.8 µeV, a limit was set on their coupling constant to photons $<4.1 \cdot 10^{-9}$, which allowed to study a previously unexplored region in the exclusion diagram. The new limit corresponds to an improvement in the sensitivity over the previous exclusion limit by a factor of approx. 7. For ALPs with a mass of 7.2 µeV a limit on their coupling constant to photons $<4.5 \cdot 10^{-8}$ GeV^{-1} was set. In a small mass range, this is more sensitive than other purely laboratory based experiments (namely laser LSW of the first generation like ALPS-1) but less sensitive than extraterrestrial experiments like CAST. For the first time ALPs have been probed by a microwave based LSW experiment.

This PhD thesis is the result of the collaboration between the CERN Beams Department in Geneva, Switzerland, and the Institute for Pulsed Power and Microwave Technology (IHM) at the Karlsruhe Institute of Technology (KIT), Karlsruhe, Germany. It has been supported by the Wolfgang-Gentner-Program of the Bundesministerium für Bildung und Forschung (BMBF).

The CERN Resonant Weakly Interacting Sub-eV Particle Search (CROWS)

Zur Erlangung des akademischen Grades eines

DOKTOR-INGENIEURS

an der Fakultät für
Elektrotechnik und Informationstechnik
des Karlsruher Institut für Technologie (KIT)

genehmigte

DISSERTATION

von

M. Eng. Michael Betz

geb. in Reutlingen

Tag der mündl. Prüfung: 17.02.2014

Hauptreferent: Prof. Dr. rer. nat. Dr. h.c. Manfred Thumm

Korreferent: Prof. Dr. Oliver Keith Baker

Eingereicht am: 10.01.2014

Abstract

The subject of this thesis is the design, implementation and first results of the "CERN Resonant WISP Search" (CROWS) experiment, which probes the existence of Weakly Interacting Sub-eV Particles (WISPs) using microwave techniques. Axion Like Particles and Hidden Sector Photons are two well motivated members of the WISP family. Their existence could reveal the composition of cold dark matter in the universe and explain a large number of astrophysical phenomena. Particularly, the discovery of an axion would solve a long standing issue in the standard model, known as the "strong CP problem". Despite their strong theoretical motivation, the hypothetical particles have not been observed in any experiment so far.

One way to probe the existence of WISPs is to exploit their interaction with photons in a "light shining through the wall" experiment. A laser beam is guided through a strong magnetic field in the "emitting region" of the experiment. This provides photons, which can convert into hypothetical Axion Like Particles or other WISPs. The photons are blocked by a wall but the WISPs practically do not interact with matter and can hence propagate to the "detection region". They can reconvert into photons, which subsequently can be detected. A hypothetical WISP would reveal itself as light, apparently shining through the opaque wall. The effect is not explained by classical physics and persists, even with flawless electromagnetic shielding. These kind of experiments have already been carried out at several institutes in the optical regime. To improve sensitivity, it has been suggested to implement Fabry-Pérot resonators on the emitting and detecting side of the experiment.

The CROWS experiment is based on the same principles, but operates in the microwave regime at a frequency of several GHz, replacing the laser by a signal generator and the optical detector by a sensitive receiver. CROWS exploits the fact, that microwave cavity resonators are technologically easier to implement than optical Fabry-Pérot ones. On the other hand, electromagnetic shielding between the fields in the two cavities is one of the primary concerns to achieve a sufficient detection sensitivity. An attenuation factor in the order of 300 dB is required to reject direct leakage below the detection threshold. The shielding has been organized in several, demountable shells, providing ≈ 100 dB of attenuation each. This allowed each shell to be characterized by conventional measurement techniques. Moreover, a test signal was emitted within the shielding to ensure that its performance did not deteriorate during the experimental run e.g., due to vibration or corrosion. Two shielding enclosures have been built, one for the detecting cavity, which was placed within a 3 T magnet and one for the components of the signal receiver, which was placed outside the magnet. Transmission of analog signals between the two shielding enclosures has been accomplished by optical transceivers, which had to be customized to operate in the strong magnetic field environment. A custom made double heterodyne signal receiver and a commercial signal analyzer have been utilized to record a 20 kHz or 2 kHz wide slice of the signal spectrum, where a WISP signal would have been expected. The power spectrum of the recorded time-trace has been evaluated, which corresponds to narrowband filtering. For a 29 h measurement run, resolution bandwidths of < 10 μHz have been achieved. This allows to detect signals with less than -210 dBm of power, corresponding to a flux of ≈ 0.5 photons/s for a 3 GHz signal. Both receivers were optimized for minimum frequency drifts by phase locking all oscillators to a common reference clock.

No HSPs or ALPs were observed in the most sensitive measurement-runs of the CROWS experiment. For HSPs with a mass of $m_{\gamma'} = 10.8$ μeV, a limit was set on the coupling constant of $\chi < 4.1 \cdot 10^{-9}$, which allowed us to explore a previously unexplored region in the parameter space. The result corresponds to an improvement in sensitivity over the previous exclusion limit by a factor of ≈ 7. For ALPs with a mass of $m_a = 7.2$ μeV we set a limit on their coupling constant of $g < 4.5 \cdot 10^{-8}$ GeV^{-1}. In a small mass range, this is more sensitive than other purely laboratory based experiments (namely laser LSW of the first generation like ALPS-1) but less sensitive than extraterrestrial experiments like CAST. This is the first time ALPs have been probed by a microwave based LSW experiment.

Kurzfassung

Die vorliegende Arbeit beschreibt die Entwicklung, Umsetzung und ersten Resultate des "CERN Resonant WISP Search" (CROWS) Experiments, welches die Existenz von schwach wechselwirkenden und leichten Teilchen (WISPs) im Mikrowellenbereich untersucht. Zwei bekannte Mitglieder der WISP Familie sind Axion Like Particles (ALPs) und Hidden Sector Photons (HSPs). Ihr Nachweis könnte die Zusammensetzung der kalten dunklen Materie in unserem Universum enthüllen und eine große Zahl astro-physikalischer Phänomene erklären. Insbesondere die Existenz eines Axions würde ein langjähriges Problem des Standardmodells lösen, welches im allgemeinen als "strong CP problem" bekannt ist. Allerdings ist der experimentelle Nachweis dieser hypothetischen Teilchen bisher nicht gelungen.

Eine Möglichkeit, diese schwer messbaren Teilchen indirekt nachzuweisen, ergibt sich durch ihre Wechselwirkung mit Photonen, welche in einem so genannten "Light Shining through the Wall" (LSW) Experiment ausgenutzt wird. Dieses besteht aus einem "Sendebereich", in dem sich ein Laserstrahl in einem starken Magnetfeld ausbreitet. In solch einer Umgebung können sich die Photonen des Laserstrahl in ALPs oder andere WISPs umwandeln. Die Photonen werden durch eine Wand gestoppt, während die WISPs praktisch nicht mit Materie interagieren und sich so in den "Empfangs-bereich" ausbreiten können. Dort wandeln sie sich, ebenfalls in einem Magnetfeld, zurück in "sichtbare" Photonen, welche anschließend detektiert werden können. Die Existenz von WISPs würde sich durch einen schwachen Lichtstrahl äußern, der scheinbar durch die undurchsichtige Wand scheint. Der Effekt ist nicht von klassisch physikalischem Ursprung und tritt auch dann auf, wenn Sende- und Empfangsbereich perfekt voneinander elek-tromagnetisch abgeschirmt sind. Um die Sensitivität des Experiments zu erhöhen, wurde bereits vorgeschlagen Fabry-Pérot Resonatoren in den beiden Bereichen zu implementieren. Diese Art von optischem Experiment wurde bisher an verschiedenen Instituten durchgeführt.

Das hier untersuchte CROWS Experiment basiert auf dem gleichen Verfahren, verwendet allerdings Mikrowellen im GHz Bereich. Der Laser wird durch eine Signalquelle und der optische Detektor durch einen sensitiven Empfänger ersetzt. Der Vorteil liegt in der technologisch einfacheren Implementierung von Resonanzstrukturen im Mikrowellenbereich (in der Form von Kavitäten) im Vergleich zu optischen Fabry-Pérot Resonatoren. Mit Mikrowellen ist die elektromagnetische Schirmung zwischen den beiden Kavitäten einer der wichtigsten Aspekte um eine ausreichend hohe Nachweisempfindlichkeit zu erreichen. Um systematische Fehler ausreichend zu unterdrücken, streben wir eine Abschirmungs-Dämpfung in der Größenordnung von ≈ 300 dB an. Dies kann nicht durch eine einzelne (demontierbare) Abschirmhülle erreicht werden sondern – in kontrollierbarer Weise – nur durch einen Mehrschalenaufbau. Dabei kann mit konventionellen Messmethoden die Hochfrequenzabschirmung jeder einzelnen Schale von etwa 100 dB gut gemessen und verifiziert werden. Es muss jedoch auch sichergestellt und bewiesen werden, dass diese Abschirmwerte während der ganzen Dauer des Experiments eingehalten und nicht etwa durch Vibration, Korrosion oder andere Einflüsse verschlechtert werden.

Im Messaufbau werden zwei Abschirmhüllen eingesetzt: Eine umschließt die Empfangskavität und muss für ALP Messungen in einem 3 T Magneten platziert werden, die andere umschließt die Komponenten des Messempfängers, der außerhalb des Magneten platziert ist. Analoge Signale werden zwischen den beiden Komponenten über eine Glasfaserverbindung übertragen. Die kommerziellen optischen Sende- und Empfangsbaugruppen mussten teilweise modifiziert werden um im starken Magnetfeld zuverlässig zu arbeiten. Das Signal der Empfangskavität wird durch weißes thermisches Rauschen dominiert. Ein WISP Kandidat würde als schwache, sinusförmige Komponente bei einer bekannten Frequenz in Erscheinung treten. Um einen 20 kHz bzw. 2 kHz breiten Bereich des Spektrums aus dem relevante Frequenzbereich zu isolieren, wurde ein selbstentwickelter Mehrfachüberlagerungsempfänger und ein kommerzieller Vector Signal Analyzer verwendet. Das Leistungsspektrum der aufgezeichneten Zeitspur wurde ausgewertet, was einer schmalbandigen Filterung entspricht. Für eine 29 h lange Aufnahme wurde eine Auflösungsbandbreite von $< 10\ \mu$Hz erreicht, was es erlaubt, Signale mit einer Leistung von < -210 dBm zu detektieren. Dies entspricht einem Fluss von ≈ 0.5 Photonen/s für ein 3 GHz Signal. Beide Empfänger wurden auf einen minimalen Frequenz-Drift hin optimiert, was durch Synchronisation aller involvierter Oszillatoren mit einem gemeinsamen Referenztakt erreicht wurde.

Keine HSP oder ALP Kandidaten konnten in den empfindlichsten Messungen des CROWS Experiment identifiziert werden. Für HSPs mit einer Masse von $m_{\gamma'} = 10.8\ \mu$eV konnte ein neues Limit für die Kopplungskonstante zu Photonen von $\chi < 4.1 \cdot 10^{-9}$ gesetzt werden, was es erlaubte einen bisher unerforschten Bereich im Parameterraum zu untersuchen. Das Ergebnis entspricht einer Verbesserung der Empfindlichkeit um den Faktor ≈ 7, im Vergleich zur bisherigen Ausschlussgrenze. Für ALPs mit einer Masse von $m_a = 7.2\ \mu$eV konnte ein Limit für die Kopplungskonstante von $g < 4.5 \cdot 10^{-8}$ GeV^{-1} gesetzt werden. Dies ist in einem kleinen Massenbereich empfindlicher als andere, rein laborgestützte Experimente (z.B. Laser LSW der ersten Generation wie ALPS-1), aber weniger empfindlich als Astronomische Experimente wie CAST. Mit dem CROWS Experiment wurden erstmals ALPs nach dem LSW Prinzip mit Mikrowellen gesucht und damit der Grundstein für weitere Entwicklungen in diesem Bereich gelegt.

List of Abbreviations and Symbols

Abbreviations

ADC	Analog to digital converter
ADMX	The Axion Dark Matter eXperiment at the University of Washington is searching for background halo axions under the assumption that they constitute the dark matter
ALP	Axion Like Particle
ALPS-1	The Any Light Particle Search – 1 at DESY is the most sensitive optical light shining through the wall experiment, searching for various WISPs
BFRT	The Brookhaven-Fermilab-Rutherford-Triests collaboration performed one of the pioneering optical light shining through the wall experiments
CASCADE	CAvity Search for Coupling of A Dark sEctor. A microwave based light shining through the wall experiment at the Cockcroft institute (UK)
CAST	CERN Axion Solar Telescope. A helioscope, searching for axions produced in the sun
CERN	European Organization for Nuclear Research
CP	Combined Charge and Parity symmetry
CPT	Combined Charge, Parity and Time symmetry
CROWS	CERN Resonant WISP search
CuBe	Beryllium copper, an advantageous material for RF gaskets and contact springs
DESY	Deutsches Elektronen Synchrotron

DFSZ	Dine Fischler Srednicki and Zhitnitski, inventors of a particular axion model
DFT	Discrete Fourier Transform
EM	Electromagnetic
EMI	Electromagnetic Interference
FET	Field Effect Transistor
FFT	Fast Fourier Transform
FOM	Figure Of Merit
GammeV	Gamma to milli-eV particle search, optical light shining through the wall experiment at Fermilab
GPS	Global Positioning System
HEMT	High Electron Mobility Transistor
HSP	Hidden Sector Photon
IF	Intermediate Frequency
KSZV	Kim Shiftman Vainshtein and Zakharov, , inventors of a particular axion model
LHC	Large Hadron Collider
LNA	Low Noise Amplifier
LNB	Low Noise Block, frontend electronics in a RF receiving chain
LO	Local Oscillator
LSW	Light Shining through the Wall
OSQAR	Optical Search for QED Vacuum Birefringence, Axions and Photon Regeneration. An optical LSW experiment at CERN
PDF	Probability Density Function
PLL	Phase Locked Loop
QCD	Quantum Chromodynamics, the part of the standard model describing interactions by the strong force
RAM	Random Access Memory
RF	Radio Frequency
SM	The Standard Model of particle physics
SPS	Super Proton Synchrotron, a particle accelerator at CERN
SQUID	Superconducting Quantum Interference Device
TE	Transverse Electric waveguide or cavity modes
TM	Transverse Magnetic waveguide or cavity modes
USB	Universal Serial Bus

UWA	University of Western Australia, they are performing a microwave light shining through the wall experiment, searching for HSPs
VNA	Vector Network Analyzer
VSA	Vector Spectrum Analyzer
WIMP	Weakly Interacting Massive Particle, a potential dark matter candidate
WISP	Weakly Interacting sub-eV Particle, a potential dark matter candidate
YMCE	Yale Microwave Cavity Experiment, microwave light shining through the wall experiment, searching for ALPs and HSPs

Symbols

B_0	Strength of the static magnetic bias field [T]
$\mathrm{BW_{res}}$	Resolution Bandwidth, the equivalent noise bandwidth of each spectral bin in a spectrum [Hz]
χ	Coupling constant between photons and HSPs [1]
ENR	Excess Noise Ratio [dB]
F	Linear noise factor [1]
f_{res}	Cavity resonance frequency [Hz]
f_{sys}	Emitting cavity excitation frequency [Hz]
g	Coupling constant between photons and ALPs [GeV^{-1}]
G	Gain [dB]
H_0	Null hypothesis in a statistical hypothesis test
H_1	Alternative hypothesis in a statistical hypothesis test
HTF	Hardware transfer function [dB]
k	Number of averaged sub-spectra
l	Linear dimension [m] or time length [s]
m_{ALP}	Mass of the ALP particle [eV/c^2]
m_{HSP}	Mass of the HSP particle [eV/c^2]
N_0	Noise power density [dBm/Hz]
NF	Noise figure [dB]
N	Number of spectral bins in a spectrum
P_{inc}	Cavity incident RF power [W]

P_{refl} — Cavity reflected RF power [W]

P_{sig} — Smallest detectable signal power [dBm]

P_N — Noise power [dBm]

PR_α — Probability for an error of the first kind (false positive) in a statistical hypothesis test

PR_β — Probability for an error of the second kind (false negative) in a statistical hypothesis test

Q_0 — Cavity unloaded quality factor [1]

Q_L — Cavity loaded quality factor [1]

S_{11} — Scattering parameter related to the normalized reflected power on a network port [dB]

S_{21} — Scattering parameter related to the normalized transmitted power between two network ports [dB]

SE — Screening effectiveness [dB]

σ_a — Allan variance [1]

SP — Span, the maximum range of frequencies which are visible in a spectrum [Hz]

θ — Scalar ALP field

T_N — Noise temperature [k]

V' — Volume of detecting cavity [m^3]

V — Volume of emitting cavity [m^3]

Contents

1. Introduction

The Standard Model (SM) in particle physics describes the interaction between matter and the fundamental forces. In a single equation, it summarizes a large part of what we know about the rules governing the world we live in. Arriving at this formulation was one of the outstanding cultural achievements of mankind, building on the work of thousands of physicists over hundreds of years. After its establishment in the 1970s, it stood the test of time, predicted many of the phenomena which have later been discovered by experiments and showed full consistency with all accepted experimental results in particle physics.

The SM is a rigid, mathematical framework, explaining how matter is made up of elementary particles, which can be divided into leptons and six different types of quarks. Combining three quarks can form particles like the proton or neutron. The family of leptons includes the electron, muon, tau and their respective neutrino counterparts. These matter particles interact with each other by four elementary forces, each one significant on a different scale. The **electromagnetic force** is the most important on a macroscopic scale – without it, we would not be able to see or even touch any object. Furthermore, in the atomic regime, it holds electrons and nuclei together. The **weak force** is very similar to the electromagnetic interaction – however, it is only relevant on a subatomic scale as its force carriers have an extremely short lifetime. The weak force is the only interaction mechanism for neutrinos, making them very weakly interacting and hence hard to detect. Moreover, it is also relevant for beta decay of certain unstable nuclei. The **strong force** holds quarks and atomic nuclei together. Compared to electromagnetism, it is about 100 times stronger but only acts over subatomic distances.

Between different pieces of matter, these forces are exchanged by carrier particles, so called bosons. The electromagnetic interaction is mediated by the photon. The weak force by the W and Z boson. The strong force by the gluon. Gravity is carried by the hypothetical graviton, which has not been detected yet and is not a part of the SM.

The mass of an elementary particle originates from its interaction with the so called Higgs field. The Higgs boson is the associated force carrier. For example, photons do not have a rest mass as they do not interact with and do not couple to the Higgs field. On 4 July 2012 the ATLAS and CMS experiments at CERN published the discovery of a particle which has been confirmed [1] to be the elusive Higgs boson, verifying yet another part of the SM. For their theoretical work which led to this discovery, François Englert and Peter Higgs [2] were awarded the 2013 Nobel prize in physics.

While the SM is our best theory to date describing all the observed phenomena related to the strong, weak and electromagnetic interaction, evidence is mounting that it is at best only an approximation to reality. A short list of its major problems is given here:

Gravity: Integrating gravitational effects in the standard model is problematic, as the most popular theory of gravity – general relativity – is not compatible with the current formulation of the SM.

Dark matter: Cosmological observations indicate, that the matter which can be explained by the SM (baryons), makes up only 5 % of the mass in our universe. The remainder are two unknown mysterious substance which are commonly called dark matter and dark energy.

Antimatter: Matter and antimatter should have been produced in equal parts in the beginning of our universe, but nowadays, antimatter is extremely rare. The SM can only partially explain this matter-abundance.

In addition to these unexplained phenomena, the SM suffers from some theoretical discrepancies. One of them is the so called "strong CP problem", which is the historical origin of axions.

1.1. The strong CP problem

Symmetries are an important feature in the SM and the theoretical issue which gave rise to the postulated existence of the axion is related to these symmetries. In physics, a symmetry corresponds to a conservation law. Classical physics and every-day life implicitly follow a space-time symmetry. Taking the collision of billiard balls as an example: the physics behind makes sense, even if we observed the table through a mirror (parity symmetry) or if a video of the shot was recorded and played backwards (time symmetry). In these observations, electromagnetism is the dominant force, for which parity (P), time (T) and furthermore charge (C) symmetry have been found to apply separately. For the sake of completeness: charge symmetry corresponds to playing billiards on a table made of antimatter.

This world view has changed radically in 1957, when it was found that the newly discovered weak force does violate parity symmetry [3]. The discovery lead to the conclusion, that the three individual symmetries do not comply with nature and that instead a combination of C and P is always conserved [4]. James Cronin and Val Fitch experimentally showed in 1965, that CP symmetry does not apply for the weak force and that hence CP is not the one true symmetry of nature [5]. This result was a total surprise at that time and 15 years later they were awarded the Nobel prize for their discovery [6, 7]. The underlying cause for the CP violation was found in the coupling of quarks to the Higgs field [8] and subsequently included in the theory of electroweak interactions. While each symmetry by itself and even the combination of CP does not apply, all observations to date and also a theoretical argument [9] indicate, that the combination of charge, parity and time symmetry (CPT) is always obeyed in the SM. A more in-depth review of symmetry violations in the SM is given in [8, 10].

The strong force is described by the theoretical framework of Quantum Chromodynamics (QCD). For several, well motivated reasons, it explicitly supports the violation of CP symmetry, in a similar way to the weak force. One argument is the apparent dominance of matter over antimatter in our universe, which can only be explained if there was significant CP violation during the Big Bang. While the weak force does break the symmetry and can account for a small part of the matter dominance, there must have been another CP violating effect, possibly from the strong force. Moreover, there are strong theoretical arguments for CP violation, namely the $U(1)_A$ problem and its solution, the Adler-Bell-Jackiw anomaly, which shall not be explained here in detail – but the interested reader can find an in-depth review in [11, 12].

For those reasons, the QCD Lagrangian contains an explicitly CP violating parameter θ. From a theoretical point of view, there is no preference for any value of θ within the range of $0 \ldots 2\pi$. If it is larger than zero and not an integer multiple of π, the strong force would be CP violating, like expected from theory. One consequence of this would be that the neutron acquires a non-zero electric dipole moment ($|d_n|$). This relation has been used to derive an experimental upper limit on the θ parameter. At the time of writing, the most sensitive experiment has been carried out at the Institut Laue Langevin in Grenoble, where polarized ultracold neutrons were used to determine d_n with high precision [13]. An upper limit was set on $|d_n| < 2.9 \cdot 10^{-26}$ e cm (with a 90 % confidence level), which relates to θ by [14]:

$$\theta = \frac{|d_n|}{5.2 \cdot 10^{-16} \ e \ cm} < 5.6 \cdot 10^{-11}. \tag{1.1}$$

It becomes obvious that the experimental limit on the CP violating parameter is over 10 orders of magnitude below the value which would have been expected. Furthermore, θ consists of two independent contributions from the strong and the weak force, which must either both be very small or of opposite sign and hence cancel each other out. While such a small value of θ is in principle allowed by theory, it does not seem like a reasonable answer as an universe with CP violation ($\theta \neq 0$) is just as likely as one without [15]. Furthermore, this would be a very unsatisfying answer for physicists, as it does not provide any new insights.

The standard model only provides a very inelegant description of nature at this point. It gives no explanation why θ is so close to zero. Alternatively, assuming that θ is equal to zero, there would be no CP violation for the strong force and θ would be an redundant parameter in the QCD theory. This corresponds to a fine tuning issue, which is commonly known in particle physics as the "strong CP problem".

A proposed solution to this problem naturally leads to an argument for the existence of axions.

1.2. A solutions beyond the Standard Model: Axions

The simplest extension to the SM, providing a solution to the problem of CP-conversation in the strong interaction was first proposed by Helen Quinn and Roberto Peccei in 1977 [16–18]. Fine tuning problems often indicate the existence of additional symmetries, which are not captured by theory. In this sense, Peccei and Quinn introduced an additional hidden symmetry[1] in the standard model. They exploited the fact that the SM is invariant under certain transformations and hence the new symmetry does not have an impact on other areas of physics. The new symmetry promoted θ from a static parameter to a time dependent variable, which can self-adjust towards zero and in this way, solve the strong CP problem very elegantly. Quinn and Peccei received the 2013 Sakurai Prize for their work on this subject.

If the new symmetry exists, it would have been spontaneously broken during the early universe, leading to CP violation and hence could explain the excess of matter over antimatter. At a later point in time (≈ 100 ns after the Big Bang), the energy density fell below the symmetry breaking scale f_a (which exact value is unknown) and θ was set to an arbitrary value, from which it started to oscillate. As there is some energy loss dampening the oscillation, θ was slowly self adjusting towards zero, explaining the non observation of CP violation nowadays[2]. Weinberg and Wilczek later found out that this dynamic oscillation can be interpreted as a field and concluded, that the Peccei Quinn mechanism inevitably leads to the existence of a very light and unusual new particle [19, 20]. Wilczek named it after an US washing detergent, emphasizing the intention of the axion to wash away a long standing problem (related to "axial" currents) in particle physics.

---

[1]To be more precise: a spontaneously broken global chiral U(1) symmetry

[2] The development of the Peccei Quinn symmetry and the axion has been illustrated in a colorful and educative analogy by P. Sikivie in [21].

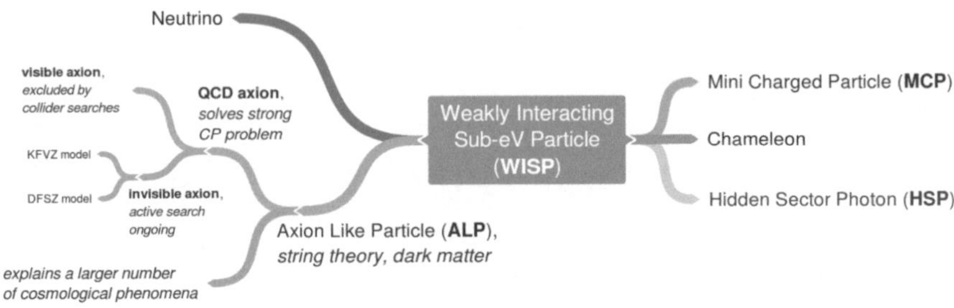

Figure 1.1.: Some notable members of the Weakly Interacting Sub-eV
Particle family.

Although alternative solutions to the strong CP problem have been postulated, the one from Peccei and Quinn is considered the most accepted
and most elegant to the present date. Nonetheless, the one remaining
obstacle consists of detecting the hypothetical particle in an experiment,
which could not be achieved within the last 30 years in spite of consider-
able efforts. Observing the axion would indicate that the Peccei Quinn
mechanism is realized in nature.

In literature this particle, which was originally proposed to solve the strong
CP problem, is often called the "QCD axion". The mass of the QCD
axion (m_a) and its coupling to two photons (g) is inversely proportional
to the symmetry breaking scale (f_a). Therefore m_a and g are in a fixed
relationship with each other, which restricts the parameter space where
the QCD axion could exist to a narrow band in the exclusion plot. For
the originally proposed QCD axion, it was assumed that f_a was on the
order of the electroweak scale (≈ 250 GeV), giving it a rather large mass
of $m_a \approx 200$ keV. This kind of axion would have been easy to detect in
collider searches and hence was called the "visible axion". It was quickly
ruled out by experiments and astrophysical arguments [17].

Nonetheless, an invisible version of the axion could still exist for $f_a >$
10^5 GeV, which corresponds to a mass in the sub-eV range and very feeble
coupling constants. Two models are popular to describe these invisible
axions, the KSVZ [22] and the DFSZ[3] model [23]. Both provide a solution of

the strong CP-problem and differ by the coupling of the axion to electrons. They are both (by design) excellent dark matter candidates and there are several active experiments, currently searching for them. One example is the Axion Dark Matter eXperiment (ADMX), which will be briefly discussed in 1.5.1. The relation between the different particle families is illustrated in Fig. 1.1.

1.3. The family of Weakly Interacting Sub-eV Particles

The axion belongs to the family of Weakly Interacting sub-eV Particles (WISPs). While it originally has been proposed as a solution to the strong CP problem, similar particles arise naturally in several extensions of the SM (particularly string theory [24]). Therefore, the axion, as it has been described so far, has been generalized to an Axion Like Particle (ALP) [25–27]. It is defined as a sub-eV mass particle with spin-zero and a very weak coupling to photons. The coupling only takes place in a strong electric or magnetic bias field and is commonly referred to as "Primakoff effect" [28]. The bias field is expressed as a "virtual photon" from a particle physics point of view, which can interact with an ordinary photon to produce an ALP. This coupling is shown in Fig. 1.2. On the other hand, ALPs can convert into photons, if a bias field is present. Up to now, all laboratory experiments searching for ALPs are based on this Primakoff effect. The main difference between the axion and the ALP is, that for the latter, the mass and coupling constant are completely independent of each other. An ALP could hence exist with an arbitrary combination of m_a and g, opening up the entire parameter space of the exclusion plot for experimental searches. The price for this generalization is, that the discovery of an ALP would not necessarily solve the strong CP problem.

Another prominent member of the WISP family is the Hidden Sector Photon (HSP) or sometimes called paraphoton, which has been postulated in [29]. The HSP is formally described by extra U(1) gauge factors in the standard model and is a generic feature of field or string theoretical SM

---

[3] Both models have been named after their inventors: Kim, Shifman, Vainshtein, Zakharov (KSVZ) and Dine, Fischler, Srednicki, Zhitnitski (DFSZ).

Figure 1.2.: Two photons (one of them being a "virtual" photon) can convert into an axion like particle and vice versa. This conversion is commonly referred to as "Primakoff effect". Virtual photons in particle physics correspond to a strong and time-constant electric or magnetic field. It is more practical to realize the latter in a laboratory environment.

extensions [26, 30, 31]. Its name has been chosen to emphasize the fact, that if it exists, it would practically not be visible to us in everyday life due to its weak interaction with all particles from the standard model and hence only exist in a so called "hidden sector". The HSP shares many of its properties with ordinary photons – but in contrast to them, possesses a small mass. The HSP couples to an ordinary photon by kinetic mixing, a process similar to neutrino flavour oscillations. That means, a photon propagating in vacuum can suddenly disappear and travel as HSP – at a later point in time, the HSP might convert back and reappear as photon. In contrast to ALPs, no electric or magnetic bias field is necessary for this process to happen. More details on HSPs and their kinetic mixing mechanism are given in [32].

Some further members of the WISP family are "Minicharged particles" [33], which are basically HSPs carrying a charge or "Chameleons" [34], which are WISPs with a variable mass, depending on the matter density of their surroundings. A comprehensive review on WISPs can be found in [26, 35].

1.4. Further arguments for the existence of WISPs

In the previous chapter, the motivation for WISPs has been shown from a more theoretical point of view. In addition there are several astrophysical phenomena, which could be explained by WISPs.

1.4.1. Dark matter

Although the axion has been postulated for a different reason, it has later been found, that the particle (among the HSP and several other WISPs) would make an excellent candidate for cold dark matter in our universe. Astronomers found evidence that only 5 % of the mass in our galaxy is due to the observable matter we know. The other 95 % is made up of two invisible substances which were given the names "dark matter" and "dark energy". The former consists of one or several unknown massive particles, which interact very little except through gravitational forces. The latter is even less tangible and thought of an uniformly distributed field in space, giving rise to a repellent force which counteracts the effects of gravity and makes the universe expand. An up-to date review on this topic can be found in [36]. The evidence for dark matter comes from the rotational velocity of objects at different distances in our galaxy, which show a disagreement with expectations. Another piece of evidence is the distortion of light from distant galaxies by gravitational effects. This so called "gravitational lensing" was observed too strong, indicating that some objects are more massive than they were expected to be [37].

It was found that the axion and the hidden photon would both make excellent candidates for the so called cold dark matter [38]. They could have been produced with a low mass and in a large number by a process called "vacuum misalignment", which happened in the early universe. Due to their weak interaction with SM particles, they would have survived until the present day, where they now make up part or all of the substance which we call dark matter.

Other candidates for dark matter include Weakly Interacting Massive Particles (WIMPS) or Massive Compact Halo Objects (MACHOS).

1.4.2. Hidden energy loss

There are other astrophysical phenomena which the SM cannot explain. For example, very high energetic ($>$ TeV) gamma rays were detected on earth. It was possible to trace back their origin to known sources, far outside the solar system. These particles were expected to interact with photons from the cosmic background radiation, which would subsequently absorb them over the very large distances involved. Astrophysicists are puzzled how these gamma rays could propagate to the earth through the "opaque" background. In other words: why is the universe more transparent than it is expected to be? A possible explanation proposes, that the high energetic particles have converted to WISPs and traveled most of the distance through the "hidden sector". Then they would convert back to observable particles, as they reach our solar system [39].

Another astrophysical phenomena where WISPs could play an important role is the cool-down of white dwarfs. These are essentially dead low-mass stars in their final lifecycle, unable to sustain fusion. White dwarfs cool down slowly by neutrino and photon emission. The observed cooling rates seem to fit better to theory, if WISPs are taken into account [40, 41]. This provides an indication that there are additional energy loss channels in effect, which are hidden from our view. One might think of WISPs which are generated within the white dwarf [42] and then fly off into space, leading to a loss of energy which is not visible to our observations. The loss mechanism can be compared to what happens in atomic power plants: up to 5 % (typically tens of megawatts) of their nominal thermal output power is radiated in the form of neutrinos, as a by-product of the nuclear fission process.

Another interesting phenomenon is supernova 1987a. The event was observed as a burst of radiation by several neutrino detectors more than 20 years ago. The length of this burst can give an indication whether WISPs were involved or not. A supernova is caused by the collapse of the core of a massive star, leading to a very dense object with a temperature of several 10 MeV. The density is so high that even neutrinos are trapped. Energy loss happens slowly due to the diffusion of neutrinos to the outer layers of the core. Energy loss through WISPs would happen faster as they interact less and hence can escape the dense environment more easily, which would lead to a shorter neutrino burst. Several large neutrino detectors on earth measured the length of this burst and make it possible to deduce limits on the properties of several WISPs from the observed data [43].

In conclusion, WISPs could explain a large number of astrophysical phenomena [44]. The axion and the HSP would be excellent candidates for cold dark matter. Particularly the axion has been the most accepted solution for the strong CP problem in QCD for over 30 years now. Moreover, string-theory naturally predicts a large number of ALPs and HSPs. Nonetheless, to date, there is no experimental evidence from laboratory searches yet and all efforts so far have only produced exclusion results.

1.5. Experimental WISP searches

After the "visible axion" has been ruled out by accelerator searches, P. Sikivie was the first one to tackle the "invisible axion" [45]. He proposed several experiments, which were theoretically able to probe the existence of this hidden particle and possibly other WISPs.

In contrast to the visible axion and many other particles from the standard model, it is not very efficient to search for WISPs in high energy collider experiments. Due to their weakly interacting nature and their low energy, a high luminosity and many collisions are needed to produce a single WISP, which is expensive if high energy particles are being used. The most promising way to detect WISPs is to exploit their coupling to photons in specialized high precision experiments, exploiting an extremely large number of lower energy particles. These experiments are usually based on the so called "Primakoff effect", which predicts that two ordinary photons can interact in a strong and static magnetic or electric field to produce an ALP. On the other hand, an ALP can decay into two photons in the aforementioned background field. In practice, strong magnets are preferred over strong electrical fields, as they are easier to realize. A quick introduction to some of the more popular experiments based on the Primakoff effect shall be given here.

Note that the sensitivity of an experiment is commonly expressed by bounds on the coupling constant of ALPs to photons (g) or HSPs to photons (χ). Smaller values indicate a more sensitive experiment. The coupling parameter and the mass of the hidden particle form the unknown parameter-space in which ALPs or HSPs are being searched for. Results are usually shown on a 2D exclusion plot (see Fig. 1.3), where a shaded area shows the regions of the parameter-space which have already been probed and in which the existence of the hidden particles has been excluded. Due

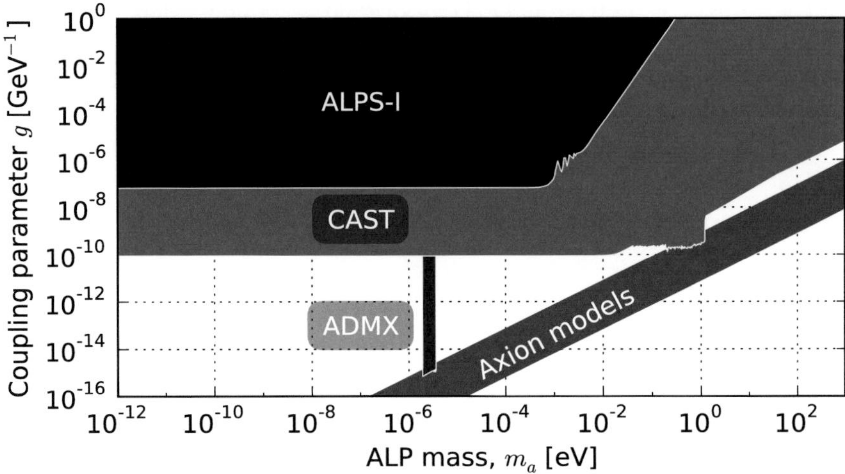

Figure 1.3.: Exclusion plot for axion like particles. Latest results from the most sensitive Laser-LSW (ALPS-1 [46]), Helioscope (CAST [47, 48]) and Haloscope (ADMX [49]) experiments are shown. The green shaded region indicates masses and coupling constants which are favoured by theory for a QCD-axion, solving the strong CP problem.

to the different nature of the experiments they cover different areas in the exclusion plot.

1.5.1. Haloscopes: WISPs from the Big Bang

For a haloscope experiment the assumption is made that so called "relic axions" have been produced copiously in the early universe and would nowadays dominate the composition of dark matter. If this is the case, one would not need to worry about the production of axions in the laboratory, as we would be constantly exposed to a large number of them (one might find trillions in a sugar cube volume [50]). The Axion Dark Matter eXperiment (ADMX) has been designed and realized at the university of Washington to detect this background dark matter. The detection volume is the bore of a 8 Tesla magnet, where some of the hypothetical axions convert to photons, which the experiment subsequently tries to detect. As the relic axions are expected at a specific energy in the range of 10^{-6} eV up to 10^{-3} eV, the reconverted photons would appear in the microwave range, corresponding to an electromagnetic field, which unexpectedly (in a sense of not being explained by the SM) appears within the detection volume. The field is "captured" by a microwave cavity, tuned to resonate with the monochromatic axion signal. A microwave receiver, optimized for extremely low background noise is used to observe this excitation of the cavity by axions.

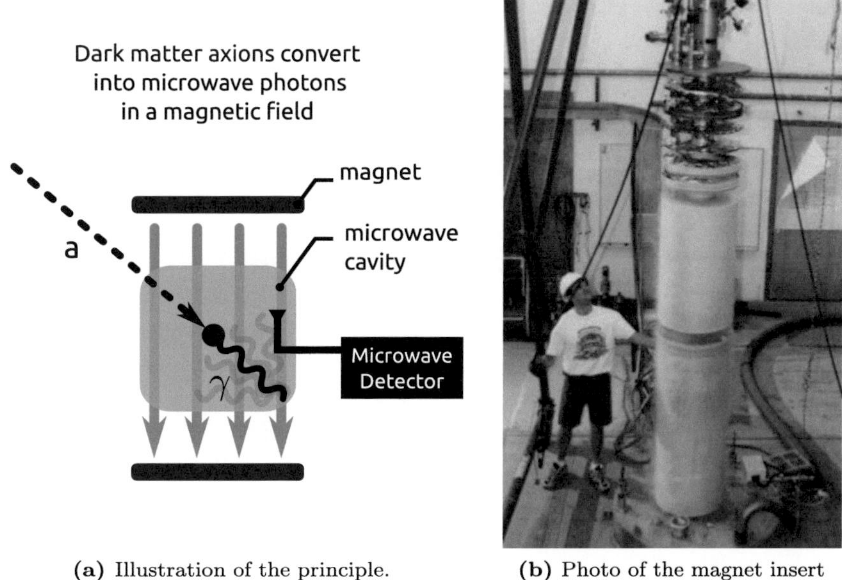

(a) Illustration of the principle.

(b) Photo of the magnet insert
©ADMX

Figure 1.4.: In the ADMX and YMCE experiments, hypothetical dark matter axions **a** can convert into microwave photons γ in the magnet, which subsequently can be detected by a sensitive receiver.

ADMX has been ongoing since 1987 and has produced the most stringent exclusion limits for QCD axions of any experiment so far. The process is slow, as the resonant cavity needs to be tuned for probing each specific axion mass. A range of less than one order of magnitude has been covered up to now [49] (by tuning the cavity from 450 MHz to 850 MHz) – it is planned to cover 3 orders of magnitude within the next 5 years.

While ADMX provides exclusion results with outstanding sensitivity (8 orders of magnitude more sensitive than current laboratory searches), these results depend strongly on the assumption that axions make up all of the dark matter and that the local dark matter density around earth is 0.45 GeV/cm^3.

Another Haloscope, searching for dark matter in a similar way to ADMX – but at a frequency of 34 GHz – is the Yale Microwave Cavity Experiment (YMCE). It will provide complementary results to ADMX, filling in the mass range around 10^{-4} eV. At the time of writing, the first measurement runs have been carried out with promising preliminary results [51, 52]. It can be expected that a new range in the exclusion plot will soon be explored by this experiment.

1.5.2. Helioscopes: WISPs from the sun

The interior of stars provides all necessary conditions to produce axion like particles. There is evidence for the existence of moving plasma currents in our closest star: the Sun. These give rise to strong electric and magnetic fields within its core, which allow the Primakoff effect to happen: Some of the internal photons convert to ALPs in the keV range. As opposed to photons, which spend several 1000 years within the suns core, ALPs don't interact with matter and could freely escape from the sun. Helioscopes like the CERN Axion Solar Telescope (CAST) try to reconvert the ALPs to detectable photons. The main part of CAST is a dipole magnet, which originally has been used as a prototype for the Large Hadron Collider (LHC). It has been mounted on a movable gantry, making it possible to point its bore at the sun for several hours each day. A hypothetical solar axion can reconvert to a photon as it propagates through the 10 m long magnet, which provides a 9 T transverse magnetic field. The reconverted photons would appear in the x-ray range and can be observed by sensitive x-ray detectors downstream of the magnet. The existence of ALPs would

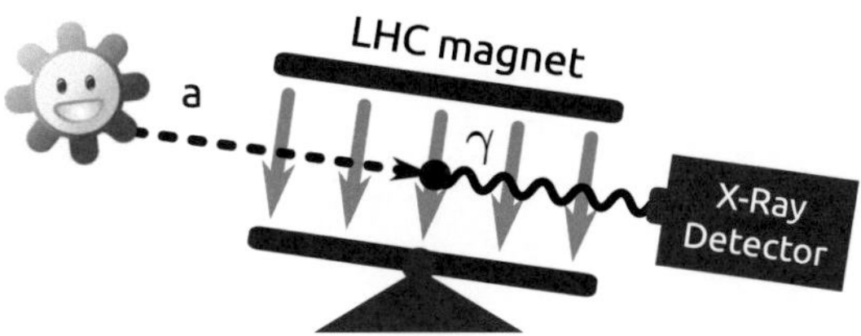

Figure 1.5.: The CERN Axion Solar Telescope makes use of a LHC dipole magnet, which is mounted on a movable gantry and pointed at the sun for several hours each day. Hypothetical ALPs from the sun **a** can convert into x-ray photons γ in the magnet. The hidden particles would indirectly appear as an excess of detected x-rays when CAST is pointed at the sun.

manifest as an excess in detected x-rays during the time when CAST is pointing at the sun. So far no ALP candidates have been identified.

CAST profits from using the sun as a strong ALP emitter and is hence about 3 orders of magnitude more sensitive than current laboratory based searches [47]. In contrast to ADMX, it is sensitive over a comparatively wide mass range from 10^{-15} eV to 10^3 eV (as shown in Fig. 1.3) and hence, at the time of writing, excludes the largest area in the parameter space of any single experiment (apart from astrophysical arguments).

1.5.3. Light shining through the wall: WISPs produced in the laboratory

So called Light Shining through the Wall (LSW) experiments search for WISPs in the laboratory without depending on any additional assumptions. They exploit the fact that WISPs can penetrate any kind of boundary and hence propagate to places which are unreachable for photons.

An experiment of this kind was first suggested by Okun in 1982 [29] and further developed into a proposal for a laboratory experiment by

Figure 1.6.: Photo of the CERN Axion Solar Telescope (CAST) ©CERN.

Sikivie, Van Bibber, Gasperini and Hoogeven [45, 53–55]. In the original proposal, a laser beam, shining through a strong magnetic field forms the "emitting region". In this environment, photons can convert into ALPs, which would propagate parallel to the photon beam. An opaque wall is placed downstream of the magnet, blocking all photons. As the ALP beam does not interact with matter, it penetrates the wall and propagates towards the "detecting region". A second magnet reconverts the ALPs to photons which can be detected by a sensitive optical instrument. To improve the low conversion probability, mirrors may be placed at each end of the emitting and detecting region, forming optical resonators and allowing the photons to pass several times through the magnetic field. The LSW principle has been illustrated in Fig. 1.7.

Examples for optical LSW experiments include GammeV [56] at Fermilab, OSQAR [57] at CERN, BMV [58] at Brookhaven Laboratory and ALPS-1 [46] at DESY. The latter is currently the most sensitive experiment of its kind. The ALPS-2 experiment [59] is currently under development, featuring a string of 10 superconducting HERA bending magnets. It will be of significantly larger scale than any of the existing LSW experiments and achieve a sensitivity, which is expected to compete with the exclusion limits from CAST.

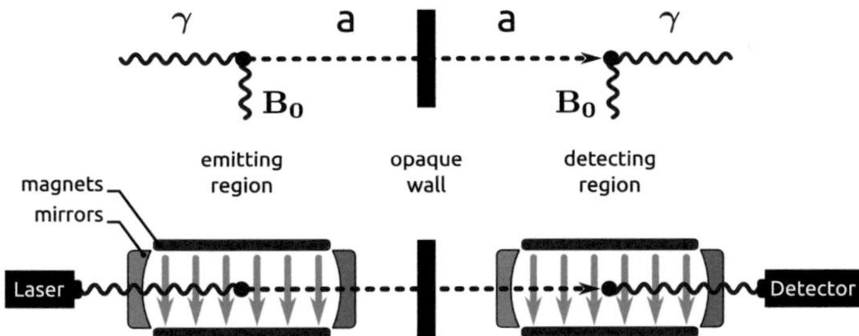

Figure 1.7.: Principle of an optical LSW experiment. Top: Photons (γ) convert into ALPs (**a**) by the Primakoff effect in a strong magnetic field (B_0), which allows them to penetrate the opaque wall. On the other side, the reverse process reconverts ALPs to photons, which can be detected. Light seems to be shining through the wall. Bottom: conversion probability can be enhanced by mirrors, which form optical cavities and reflect the light many times through the magnet.

The biggest advantage of LSW experiments is that they do not depend on models of axion flux or local axion density, hence minimizing the number of prior assumptions and producing the most unambiguous exclusion limits. Their biggest disadvantage is, that the very low interaction probability of WISPs is involved twice, once for converting photons to WISPs and again for converting WISPs back to photons. Therefore LSW experiments cannot compete with haloscope searches like ADMX, where only one conversion is necessary. On the other hand, LSW experiments have the advantage of being inherently broadband and sensitive over a wide WISP mass range. They are only restricted on the higher end of the mass scale by the energy of the photons in use (several eV in the optical regime).

Proposals have been written to adapt the LSW principle to microwaves [60–62]. This offers several advantages compared to optical light, e.g., microwave cavities are cheaper and technologically easier to implement as optical resonators.

Microwave LSW experiments have been carried out at the University of Western Australia (UWA) at 20 GHz [63], by ADMX at 700 MHz [64] and by YMCE at 34 GHz [52]. So far, all of them were searching for HSPs without a magnetic bias field and the sensitivity of each one was limited by the performance of the electromagnetic shielding.

The CROWS experiment is not suffering from these limitations and improved sensitivity by about 4 orders of magnitude compared to the UWA experiment and 1 order of magnitude compared to the ADMX-HSP experiment. We were able to exclude HSPs in a previously unexplored area in the 10 μeV mass range. CROWS is also the first microwave based LSW experiment searching for ALPs in a 3 T magnetic field and was able to confirm the ALPS-1 results.

A microwave LSW experiment, which will start its first measurement runs at 1.3 GHz in the near future is the CAvity Search for Coupling of A Dark sEctor (CASCADE) at the Cockcroft Institute [65]. While all of the previously mentioned LSW setups use copper cavities, CASCADE will be the first one to exploit the advantages of superconducting cavities for HSP search. The setup will make use of the same narrowband receiving concept, which has been pioneered by the CROWS experiment.

2. Theoretical background of LSW - experiments

One of the most promising ways to probe the existence of WISPs is to exploit their weak interaction with photons. For example, ALPs can convert into photons if a strong magnetic field is present (and vice versa). These effects would require a modification to the classical equations of electrodynamics. In 2.1, these modifications are reviewed and discussed. Subsequently, in 2.2, it is shown how this coupling from WISPs to photons can be exploited in a practical LSW experiment in the microwave range. The equations for the detection sensitivity of such an experiment are presented and analyzed in 2.3. In 2.4, the geometric factor G and its numerical determination is explained. Furthermore, the radiated HSP fields of various microwave resonators are visualized.

2.1. Modified Maxwell equations

WISPs interact very weakly with standard model particles and besides gravitational interaction, their weak coupling to photons is one of the most promising ways to indirectly observe them experimentally. Most experimental WISP searches to date rely on the coupling of an "unobservable" WISP to an observable photon. For ALPs this happens by the Primakoff effect. Axions can convert to photons and photons can convert to axions, if a strong static magnetic or electric field is present. If we only take this coupling to photons into account, we can simplify the relevant observable effects of an ALP field to an additional coupling term in Maxwells equations. These are the classical equations for static and propagating electromagnetic fields, which were derived by James Clerk Maxwell in 1861.

Note that the equations in this section are not based on SI units, but rely instead on "natural units", more specifically on the Lorentz-Heaviside system with an additional normalization of $c = 1$. This system of units has been adopted for several reasons. Foremost, because it is a popular choice in literature related to WISPs, ALPs and HSPs. Choosing this system of units allows the equations to be consistent with the ones presented in literature. Furthermore, the goal of this section is not to derive the modified Maxwell equations, which has already been done in [45, 60, 61, 66–68], or to carry out numerical calculations – but to show the phenomenology of an hypothetical ALP and its implications if it would exist, in an intuitive way. For that purpose, the choice of the system of units is not a primary concern.

The classical Maxwell equations for a homogeneous and isotropic medium and with the system of units as explained above, are given by:

$$\nabla \cdot \mathbf{E} = \tilde{\rho}, \quad \nabla \cdot \mathbf{B} = 0, \quad \nabla \times \mathbf{E} = -\frac{\partial \mathbf{B}}{\partial t}, \quad \nabla \times \mathbf{B} = \frac{\partial \mathbf{E}}{\partial t} + \tilde{j}, \quad (2.1)$$

where $\mathbf{E}$ is the electric field, $\mathbf{B}$ is the magnetic field, $\tilde{\rho}$ is the charge density and $\tilde{j}$ is the current density. If we consider free space (like in a vacuum), there is nothing which could carry an electric charge and hence nothing to support the flow of an electromagnetic current. Therefore, the equations can be simplified by setting $\tilde{\rho} = 0$ and $\tilde{j} = 0$.

So far, no WISPs have been considered and these equation agree with the standard model, which has been experimentally tested many times. Now we will modify these equations to take the weak coupling to WISPs into account.

2.1.1. Axion Like Particles

An "Axion Like Particle" is formally the carrier of a pseudoscalar field θ with a sub-eV mass. The ALPs couple weakly to the electromagnetic field by the Primakoff effect [28], which requires an external magnetic (or

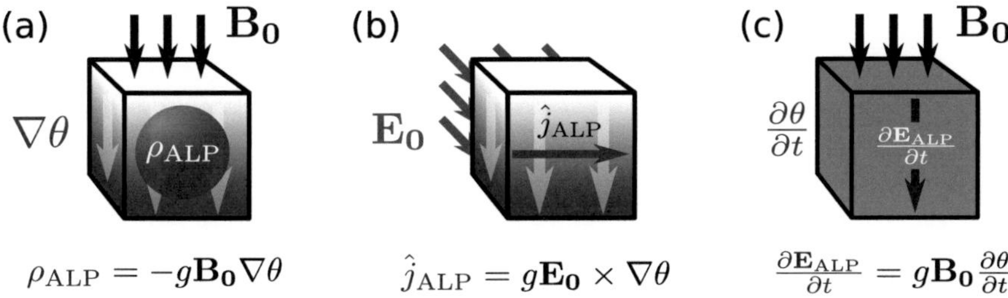

Figure 2.1.: An hypothetical ALP would interact with electromagnetic fields, leading to three observable phenomena, which cannot be explained by the traditional Maxwell equations: **(a)** an ALP field gradient leads to an observable charge in an external static magnetic field. **(b)** an ALP gradient leads to an observable current in an external electric field. **(c)** a time dependent ALP field leads to an observable time dependent electric field in an external static magnetic field.

electric) bias field. The Maxwell equations can be modified [60, 66, 69] to account for this weakly coupled pseudoscalar field:

$$\nabla \cdot \mathbf{E} = \tilde{\rho} - g\mathbf{B} \cdot \nabla\theta \tag{2.2}$$

$$\nabla \cdot \mathbf{B} = 0 \tag{2.3}$$

$$\nabla \times \mathbf{E} = -\frac{\partial\mathbf{B}}{\partial t} \tag{2.4}$$

$$\nabla \times \mathbf{B} = \frac{\partial\mathbf{E}}{\partial t} + \tilde{\jmath} + g\mathbf{B}\frac{\partial\theta}{\partial t} - g\mathbf{E} \times \nabla\theta, \tag{2.5}$$

where g [1/GeV] is a constant determining the strength of the coupling between ALPs and photons. If the ALP exists, it is due to the very small value of g, that the additional terms can not be observed in everyday life. The modification involves an additional term in Gauss' law and Amperes law: one looks just like an additional electric charge density and the other looks just like an additional electromagnetic current – however both originate from the hidden ALP. The remarkable thing is, that these two effects could even be observed in free space, in the complete absence of any ordinary charge carriers.

The additional charge density in Gauss' law (Eq. 2.2) is proportional to $g\mathbf{B} \cdot \nabla\theta$. It only becomes apparent, when there is spatial change in the ALPs field – for example by some kind of ALP boundary[1]. This has been illustrated in Fig. 2.1(a), where the box contains an ALP field with a certain gradient. If an external magnetic field is aligned parallel to this gradient, we can observe the accumulation of charge. Note that what we actually would observe, is an electric field which seems to originate from an apparent electrical charge – but in this case, there would be no charged particles involved.

It is interesting to note that this effect could be exploited in an electrostatic experiment. The electric field strength of an object with a well defined charge would have to be measured. Then the measurement would be repeated in a strong magnetic field. If a local ALPs gradient exists and if it is aligned parallel to the magnetic field lines, the $g\mathbf{B} \cdot \nabla\theta$ term would influence the apparent charge of the measured object. Rotating the magnet or the experimental inset would allow to modulate the effect.

Furthermore, Amperes law in Eq. 2.5 is modified. Two new terms are added, which both give rise to an effect, looking just like an additional electromagnetic current $\tilde{j}$. This has been illustrated in Fig. 2.1(b). A gradient in the ALPs field is needed. Applying an electric field, perpendicular to the ALPs gradient results in a flow of current, perpendicular to both. Note, that what we have called "current" is not really a current in the sense of moving electrons. It is more an observable effect caused by the hidden ALPs, which looks in every way like an ordinary current – but unlike that, can also exist in the absence of charge carriers.

The second term, modifying Amperes law, is $g\mathbf{B}\frac{\partial\theta}{\partial t}$ and has been illustrated in Fig. 2.1(c). It is different from the other two terms in the sense of being based on a time derivative instead of a spatial gradient. This makes it very interesting from an experimental point of view. Consider a time varying ALPs field $\theta(t)$, which interacts with an external static magnetic field B_0. This would give rise to an electric field, pointing in the same direction as B_0 and having the same time derivative as $\theta(t)$. The detecting cavity of the CROWS experiment is driven by this electric field, which would seemingly appear out of nowhere in the magnetic detection volume,

---

[1]Note that due to the weakly interacting nature of the ALPs, this is very hard to achieve in practice and shall only serve as an example to explain the phenomenology.

if ALPs exist. On the other hand, a strong and time varying electric field in a magnet can give rise to an ALPs field, if $\mathbf{E}$ is aligned with B_0, which is the physical process happening in the emitting cavity.

At the time of writing, the additional terms in Maxwells' equations could not yet be confirmed by observations in free space. Remarkably though, analogous effects have recently been observed in condensed matter physics [70, 71]. A certain family of crystalline materials, which are called "topological insulators", exhibit something similar to the axion field (θ). They can be seen as a realization of axion electrodynamics: On the boundary of these materials, the electromagnetic fields will behave according to the rules given by the modified Maxwell equations. Note, however, that we cannot draw any conclusion on the existence of ALPs from this, as the underlying cause for these effects is substantially different to the one searched for in particle physics, making them only observable within these very special materials and not in vacuum. In other words, an ALP requires the realization of axion electrodynamics in free space, and not in a highly complex medium.

The modified Maxwell equations, taking ALPs into account, have been solved in [60, 67]. As the ALPs only interact very weakly with matter, no boundary conditions can be enforced and a solution to the differential wave equations is given by the "retarded massive Green's function". We first define the external magnetic field as the product of an amplitude (B_0) and a spatially dependent vector field, which has been normalized to $|\mathbf{B}_n(\mathbf{x})| = 1$:

$$\mathbf{B}(\mathbf{x}) = B_0 \cdot \mathbf{B}_n(\mathbf{x}) \tag{2.6}$$

We apply the same separation for the time dependent and periodic electrical field with the angular frequency ω, which could for example originate from the resonating mode of a microwave cavity:

$$\mathbf{E}(\mathbf{x}, t) = E_0 \cdot e^{-j\omega t} \cdot \mathbf{E}_n(\mathbf{x}) \tag{2.7}$$

The time dependent ALPs field $\theta(\mathbf{x}, t)$ at the point $\mathbf{x}$ in space, can then be determined by:

$$\theta(\mathbf{x}, t) = -g \, B_0 \, E_0 \, e^{-j\omega t} \int_V \frac{e^{jk'|\mathbf{x}-\mathbf{y}|}}{4\pi \, |\mathbf{x} - \mathbf{y}|} \, \mathbf{B}_n \cdot \mathbf{E}_n \, d^3\mathbf{y}, \qquad (2.8)$$

where g is the ALP coupling constant and k' the wavenumber of the ALP, taking its non-zero rest mass (m_a) into account:

$$k' = \sqrt{\omega^2 - m_a^2}. \qquad (2.9)$$

The integration in Eq. 2.2 runs over the volume, filled by the electric field. The vector $\mathbf{y}$ indexes a point in this volume. The dot product $\mathbf{B}_n \cdot \mathbf{E}_n$ suggests that we will obtain the strongest radiated ALPs field if the electric field is parallel to the external magnetic field over a large volume. This has already been observed in Fig. 2.1(c). The field falls off proportionally with distance. The propagation phase of the wave is different from the electromagnetic one, due to the non-zero rest mass of the ALP. More details on a possible computational method to evaluate Eq. 2.8 are given in 2.4.

For the sake of completeness, we will mention the ALPs Lagrangian, which is often used in literature, as it encodes the above modified Maxwell equation in a more general and compact form:

$$\mathcal{L} = -\frac{1}{4} F_{\mu\nu} F^{\mu\nu} + \frac{1}{2} \partial_\mu \theta \partial^\mu \theta - \frac{1}{2} m_a^2 \theta^2 + \frac{1}{4} g\theta F_{\mu\nu} \tilde{F}^{\mu\nu}, \qquad (2.10)$$

where $F_{\mu\nu}$ is the electromagnetic field strength tensor and $\tilde{F}^{\mu\nu}$ is its dual.

We can make the following important conclusions on the features of ALPs:

- ALPs couple to photons (and vice versa) in a strong magnetic field $(\mathbf{B}_0)$ by the Primakoff effect.

- The ALPs field is scalar in nature and originates from $\mathbf{E} \cdot \mathbf{B}_0$, so coupling is maximized, if $\mathbf{E}$ is parallel to $\mathbf{B}_0$.

- ALPs are massive, hence propagate slower than photons. This leads to a different phase velocity of the ALP field.

- ALPs only interact very weakly with matter and hence can penetrate any boundaries, walls or obstacles.

2.1.2. Hidden Sector Photons

Another hidden particle, which is quite similar in its observable effects to the ALP is the Hidden Sector Photon. While ALPs couple to photons by the Primakoff effect, HSPs couple to photons by kinetic mixing. In practice, HSPs are so similar to ALPs that they can also be detected by a LSW experiment. In fact, the apparatus for testing the existence of HSPs would be identical to the ALPs case, the only difference is, that no magnet is required. As we have used the data from experimental runs to derive exclusion limits for ALPs and HSPs, the theoretical background for HSPs shall be briefly explained in this section.

A "Hidden Sector Photon" can formally be described by extra $U(1)$ gauge factors in the standard model. Qualitatively, the HSP field appears similar to the electromagnetic field from which it originates. It differs due to the fact that HSPs do not interact with matter. The HSP field from a cavity resonator, for example, is not stopped by the cavity walls and continues to propagate in all directions from there. In contrast to the photon, the HSP is massive. It propagates slower than the speed of light, which results in a different phase velocity. Some illustrations of how the HSP fields from the emitting cavity in the CROWS experiment look like, are given in 2.4.

The HSP equations of motions have been solved in [61]. The following equation describes the HSP field ($\mathbf{B}$) at the point $\mathbf{x}$:

$$\mathbf{B}(\mathbf{x}, t) = \chi \, m_{\mathrm{HSP}}^2 \, E_0 \, e^{-j\omega t} \int_V \frac{e^{jk'|\mathbf{x}-\mathbf{y}|}}{4\pi \, |\mathbf{x} - \mathbf{y}|} \, \mathbf{E}_n \, d^3\mathbf{y}, \qquad (2.11)$$

where χ is the HSP coupling constant due to kinetic mixing and ω the frequency of the electromagnetic field. As in Eq. 2.7, the electric field has been separated into a time dependent ($E_0 \, e^{-j\omega t}$) and a normalized spatially dependent ($\mathbf{E}_n$) part. The propagation factor $e^{jk'|\mathbf{x}-\mathbf{y}|}$ in Eq. 2.11

takes the non-zero rest mass (m_{HSP}) of the HSP into account, as it depends on the wavenumber k' of the HSP, which is defined as:

$$k' = \sqrt{\omega^2 - m_{\mathrm{HSP}}^2}.\tag{2.12}$$

The HSP Lagrangian is:

$$\mathcal{L} = -\frac{1}{4}F^{\mu\nu}F_{\mu\nu} - \frac{1}{4}B^{\mu\nu}B_{\mu\nu} - \frac{1}{2}\chi F^{\mu\nu}B_{\mu\nu} + \frac{1}{2}m_{\gamma'}^2 B^{\mu}B_{\mu}\tag{2.13}$$

where $F^{\mu\nu}$ and $B^{\mu\nu}$ describe the gauge field strengths for the photon and the HSP respectively. B_{μ} is the HSP gauge field. The first term describes the standard kinetic term for the photon or electromagnetic propagation. The second term describes HSP propagation. The third term describes the kinetic mixing between the photon and the hidden photon. The fourth term gives mass to the HSP.

We can make the following important conclusions on the features of HSPs:

- HSPs couple to photons (and vice versa) without any external driving force by "kinetic mixing".

- HSPs are described by a vector field, which appears very similar to the electromagnetic field from which they originate.

- HSPs are massive, hence propagate slower than photons. This leads to a different phase velocity of the HSP field.

- HSPs interact very weakly with matter and can penetrate any boundaries, walls or obstacles.

2.2. Principle of "Light Shining through the Wall" experiments

In 2.1 we observed that a time-varying electric field will couple to a time varying ALP or HSP field and vice versa. This effect has been exploited in so called Light Shining through the Wall (LSW) experiments in the optical range, as described in 1.5.3.

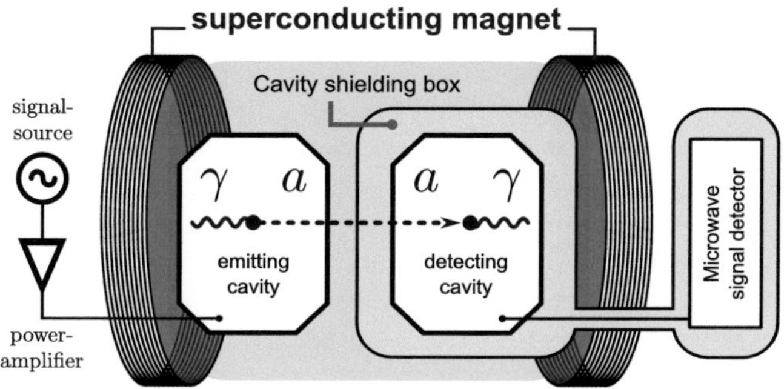

Figure 2.2.: Basic block diagram of a microwave LSW experiment.

Adapting the LSW principle to microwaves has first been proposed in [60, 61]. The main motivations for using microwaves are:

- It is technologically less challenging to produce and operate low loss resonators in the microwave than in the optical regime, rendering the experimental setup cheaper, more rugged and easier to handle. The realization of optical cavities is a substantial challenge due to the small wavelengths and therefore stringent mechanical tolerances involved.

- For microwave cavities, the separation between them can be reduced to be in the order of the involved wavelength, making the experiment sensitive – not only to propagating WISPs (like in a laser based experiment) – but also to "near field" WISPs [60] surrounding the cavities.

- The energy needed to produce photons decreases with longer wavelengths. For a given input power, more photons can be produced in microwave based experiments, they are therefore more sensitive to WISPs than optical ones. However, the lower photon energy also poses a disadvantage by restricting the maximum detectable mass of the hidden particle.

We performed a microwave based LSW experiment searching for ALPs and HSPs. The schematic of our setup is shown in Fig. 2.2.

The first ingredient for such an experiment is a strong electromagnetic field. A very effective and practical way to achieve this is a microwave cavity. It is a resonator, consisting of nothing more than a hollow metallic shell with low conduction losses to support displacement currents. If a microwave signal is coupled in at one of the resonant frequencies, the field strength within is enhanced by $\sqrt{Q}$, where Q is the quality factor and typically on the order of 10^3 - 10^5 for normal conducting technology. From a particle point of view, we can say that the emitting cavity holds a large number of microwave photons, each one with a quantum energy of:

$$E_{ph} = h \cdot f \qquad E_{ph} \approx 4.1 \ \mu eV \ \ \text{for} \ \ 1 \ GHz, \qquad (2.14)$$

where h is Planck's constant and f the excitation frequency of the cavity.

These photons couple to HSPs by kinetic mixing, or – if the cavity is placed in a strong magnet – to ALPs by the Primakoff effect. Conveniently, apart from the magnet, HSPs and ALPs can be probed by the same experimental apparatus. The WISPs easily penetrate the cavity walls and radiate outward into space. For cavities operated on a fundamental mode, there is no preferable radiation direction – however, modes of higher order (like a waveguide resonator or a laser beam) show directional WISP radiation. Some of the WISPs propagate to the detecting cavity and reconvert to photons. The cavity is tuned to resonate with this unexpectedly appearing (in the sense of not being explained by the SM) electromagnetic driving force. The excitation is measured with a sensitive microwave receiver. To prevent photons from taking the direct shortcut between the two cavities, a electromagnetic shielding enclosure is necessary.

The mass of the ALP (and also of the HSP) is a fixed but unknown parameter, which is only weakly bound by cosmological observations to the range of 10^{-12} eV $\leq m_a \leq 10^3$ eV [72]. It is important to note, that even though we emit and detect monochromatic signals of fixed frequency (and hence photons of fixed energy), the experiment is sensitive to a wide range of WISP masses. This is due to the fact that there is no energy loss associated with the conversion process: the entire energy of the photon E_{ph} converts into rest mass and into kinetic energy of the ALP or HSP. In the detecting cavity, both forms of energy convert back to a photon with the same energy as the initial one. Hence the energy of the reconverted photon is independent of WISP mass. Furthermore, the detected microwave signal will be identical in frequency to the one driving the emitting cavity. The fact that the exact frequency of the WISP signal is known has been exploited in the data evaluation and made a sensitive detection scheme possible.

2.3. Expected sensitivity of a microwave based LSW experiment

The expected output power (or photon flux) from the detecting cavity due to ALPs or HSPs has been derived in [61] and is given by:

$$P_{\mathrm{ALP}} = \left(\frac{gB_0}{f_{\mathrm{sys}}}\right)^4 |G_{\mathrm{ALP}}|^2 Q_{\mathrm{em}} Q_{\mathrm{det}} P_{\mathrm{em}}, \qquad (2.15)$$

$$P_{\mathrm{HSP}} = \chi^4 \left(\frac{m_{\gamma'}}{f_{\mathrm{sys}}}\right)^8 |G_{\mathrm{HSP}}|^2 Q_{\mathrm{em}} Q_{\mathrm{det}} P_{\mathrm{em}}, \qquad (2.16)$$

where Q_{em} and Q_{det} are the loaded quality factors of emitting and detection cavity, f_{sys} is the operation frequency of the experiment (to which the cavities are tuned) and P_{em} is the emitting cavity driving power. B_0 is the strength of the magnetic field. The unknown coupling constants for ALPs or HSPs to photons are given by g and χ respectively. Details of the geometric form factors $|G_{\mathrm{ALP}}|$ and $|G_{\mathrm{HSP}}|$ are given in 2.4. Both equations are a function of rest mass of the hidden particle, directly by $m_{\gamma'}$ (only Eq. 2.16) and indirectly by the mass dependent geometric form factors (both equations). Note that Eq. 2.15 and Eq. 2.16 are based on the system of "Natural Units". For numerical calculation, all variables must be converted to units based on [eV] or powers of it. The following conversion factors are useful for that purpose: $1\ \mathrm{T} \mathrel{\widehat{=}} 195\ \mathrm{eV}^2$, $1\ \mathrm{GHz} \mathrel{\widehat{=}} 4.135\ \mu\mathrm{eV}$.

To get a sense for the scale of the numbers involved, some example values for a realistic ALPs experiment with normal conducting cavities are:

$$P_{\mathrm{em}} = 50 \text{ W}, \quad Q_{\mathrm{em}} = Q_{\mathrm{det}} = 10000, \quad B_0 = 3 \text{ T},$$
$$f_{\mathrm{sys}} = 1.75 \text{ GHz}, \quad |G_{\mathrm{ALP}}| = 0.6, \quad g = 6 \cdot 10^{-8} \text{ GeV}^{-1} \qquad (2.17)$$
$$\Rightarrow \quad P_{\mathrm{ALP}} = -210 \text{ dBm} \approx 1 \text{ photon/s}.$$

To produce an exclusion limit, which can be compared to other experiments, P_{ALP} or P_{HSP} from Eq. 2.15 or Eq. 2.16 is set to the minimum detectable power of the microwave receiver (P_{sig}). This allows us to derive an upper bound on the coupling parameter g or χ from the other known values. If the coupling between HSPs, or ALPs and photons is stronger than this upper bound, the hidden particle would have been visible in the experiment. The sensitivity of an experiment is hence commonly expressed by bounds on g or χ, where smaller values indicate a more sensitive experiment.

The two equations (Eq. 2.15 and Eq. 2.16) show that the sensitivity for ALPs is largely dominated by the strength of the magnetic field B and the operating frequency f_{sys}, while the sensitivity for HSPs is dominated by the Q factors of the cavities and the geometric form factor $|G_{\mathrm{HSP}}|$. These parameter are therefore critical and need to be optimized for a sensitive LSW experiment.

2.4. Numerical calculation of the geometric form factor $|G|$

The sensitivity of the LSW experiment is strongly dependent on the geometry of the cavities, their size, orientation and position relative to each other and on the electric field of their resonating modes. These effects are taken into account by the geometric form factor G [60–63]. Furthermore G depends on the relative direction of the static magnetic field for ALPs and on the mass of the hidden particles.

The geometric form factor is comparable to an antenna gain, as it is commonly used in communication systems, but taking the propagation due to WISPs (having a small mass) and the full three dimensional coupling by the **near-field** into account. This is in contrast to laser LSW experiments, where the distance between the cavities is much larger than the involved wavelengths and a far field approximation is commonly used.

Formally, G corresponds to the overlap integral between the radiated WISP field (according to Eq. 2.8 or Eq. 2.11) from the emitting and detecting cavity. The larger the overlap between the two fields, the stronger the coupling between the cavities due to the hidden particles. The value of G can be determined by a 6 dimensional integration over the volumes of emitting (V) and detecting cavity (V'). The formulas, according to [61], are given for ALPs in Eq. 2.18 and for HSPs in Eq. 2.19:

$$G_{\text{ALP}} = \frac{k^2}{4\pi} \int_{V'} \int_{V} \frac{e^{ik'|\mathbf{x}-\mathbf{y}|}}{|\mathbf{x}-\mathbf{y}|} E_B(\mathbf{x}) E'_{B'}(\mathbf{y}) \, d^3\mathbf{x} d^3\mathbf{y}, \qquad (2.18)$$

$$G_{\text{HSP}} = \frac{k^2}{4\pi} \int_{V'} \int_{V} \frac{e^{ik'|\mathbf{x}-\mathbf{y}|}}{|\mathbf{x}-\mathbf{y}|} \mathbf{E}(\mathbf{x}) \cdot \mathbf{E}'(\mathbf{y}) \, d^3\mathbf{x} d^3\mathbf{y}, \qquad (2.19)$$

where $\mathbf{E}(\mathbf{x})$ is the electric field in the cavities and E_B is the component of the electric field, pointing in the same direction as the external static magnetic field. Each variable of integration, $\mathbf{x}$ and $\mathbf{y}$, represents a three dimensional vector, indexing a point within the emitting and receiving cavity on a common coordinate system. The factors k and k' are the wavenumber of the photon and the WISP, through which G becomes a function of the rest mass of the hidden particle (m_{WISP}). The relation is given by Eq. 2.20. This mass dependence has a significant influence on the shape of the excluded areas. Both integrands contain an attenuation factor proportional to distance ($|\mathbf{x}-\mathbf{y}|$) and a phase factor ($e^{ik'|\mathbf{x}-\mathbf{y}|}$), which becomes more significant with larger k' (corresponding to faster WISPs). Note that k' can become a complex number in the non-propagating WISP case (if $m_{\text{WISP}} > hf_{\text{sys}}$), leading to an exponential suppression of $|G|$.

$$k = \frac{2\pi f_{\text{sys}}}{c} \qquad \frac{k'}{k} = \sqrt{1 - \left(\frac{m_{\text{WISP}}}{hf_{\text{sys}}}\right)^2}. \qquad (2.20)$$

The $\mathbf{E}$ fields are normalized such that:

$$\int_V |\mathbf{E}(\mathbf{x})|^2 \, d^3\mathbf{x} = 1. \tag{2.21}$$

For HSPs, Eq. 2.19 suggests the dot product between the two electric fields, $\mathbf{E}(\mathbf{x}) \cdot \mathbf{E}'(\mathbf{y})$, is of importance, it is thus advantageous to use the same mode in both cavities. For ALPs Eq. 2.18 suggests that only the component of the electric field in the cavity, aligned with the static magnetic field, E_B and $E'_{B'}$, is of significance. Good geometric form factors can be expected, if the E-field in the cavity is parallel to the magnetic field lines over a large volume.

A numerical integration routine has been implemented in Python to compare $|G|_{\mathrm{HSP}}$ and $|G|_{\mathrm{ALP}}$ for different cavity modes. The results, as a function of the normalized WISP wavenumber are shown in Fig. 2.3.

The geometric form factor can be understood intuitively by visualizing the HSP field, radiated from certain cavities. This has been done by numerically evaluating Eq. 2.11 for 10^6 points in a space around the cavity. The radiated HSP vector-field has been visualized with the software package Mayavi. The result is shown in Fig. 2.4a for the TE_{011} mode in the emitting cavity.

It can be seen that the HSP field is more intense on the sides of the cavity than above or below it. This suggest a better G factor if both cavities are placed next to each other instead of being stacked. The observation was confirmed by the calculated G factor of both cases in Fig. 2.4b.

Another interesting point is to exploit the directivity of the HSP field generated by certain resonating structures. For example, a piece of waveguide can be closed by metallic endcaps, short-circuiting both ends. This forms a simple microwave resonator. The TE_{10n} mode of such an resonator has a sinusoidal electric field, pointing along the y-axis and shows n half waves along the z-direction, as illustrated in Fig. 2.5a. This mode has been analyzed for its radiation pattern of HSPs in Fig. 2.6. For a particular HSP mass, the structure shows strong directionality, radiating HSPs along its axis. For even higher order modes, this structure approaches the case of an optical LSW experiment, where the HSPs propagate in an axial beam.

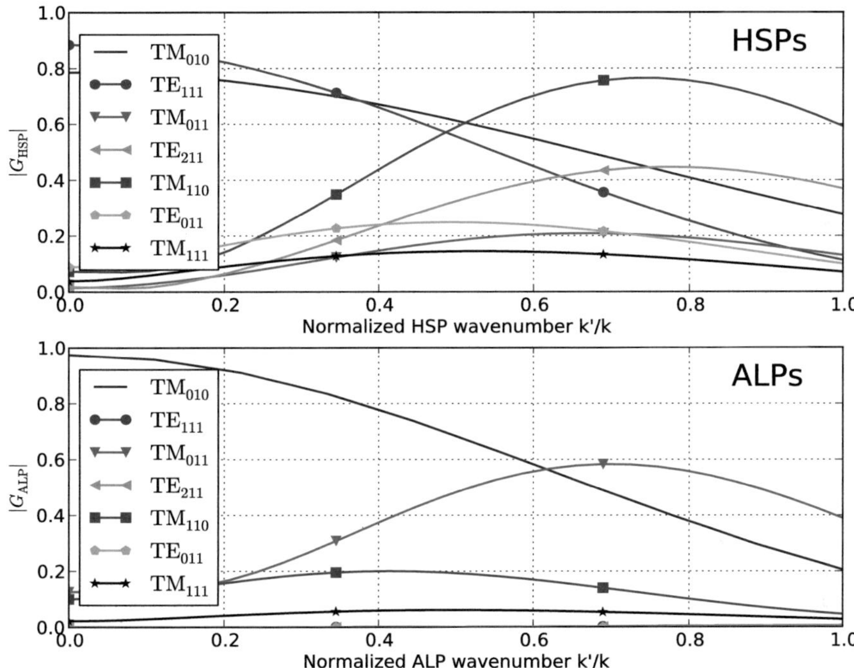

Figure 2.3.: Numerical results for the geometric form factor $|G|$ for different modes as a function of wavenumber. Cavities are aligned axial, with 20 mm distance between them. A normalized wavenumber of k'/k = 0 implies massive and slow WISPs while k'/k = 1 implies massless relativistic WISPs.

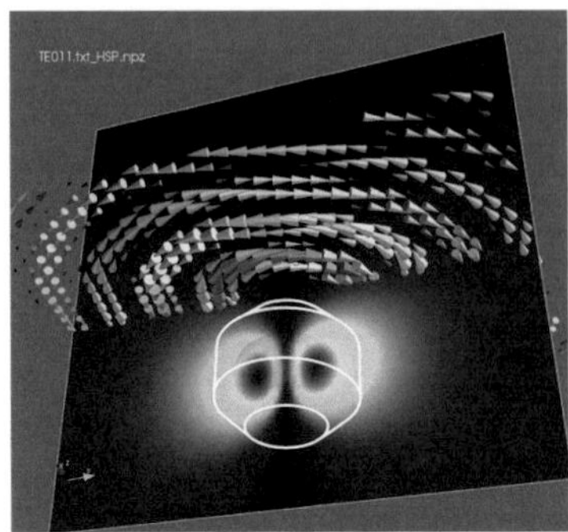

(a) HSP field generated by the TE$_{011}$ mode for k'/k=0.6

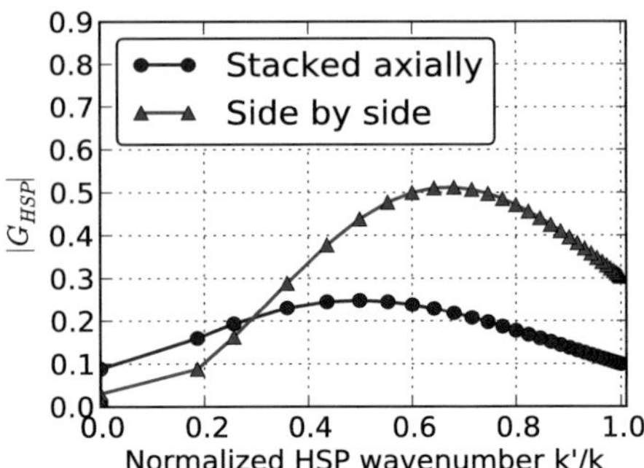

(b) Geometry factor $|G|$ of the TE$_{011}$ mode

Figure 2.4.: Intensity (colour coded) and direction (shown as cones) of the HSP field around the emitting cavity on two cutting planes. The HSP field can be compared to a continuation of the E field outside the cavity wall boundaries. Note how the field is less intense in axial direction with the cavity. This reflects the lower $|G|$ factor when placing the cavities axially on top of each other.

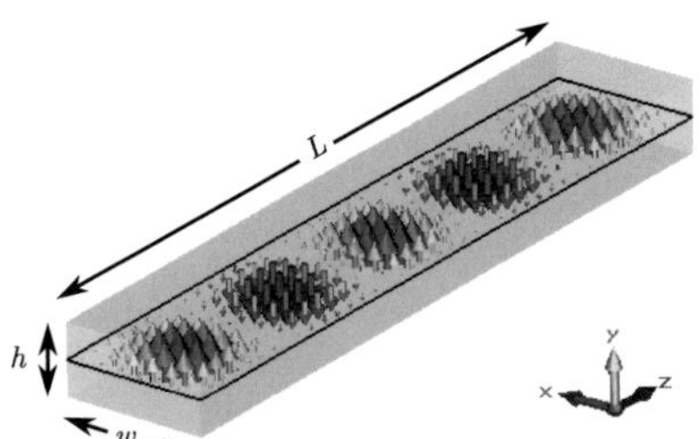

(a) Electric field configuration of the TE$_{105}$ mode in a rectangular waveguide resonator.

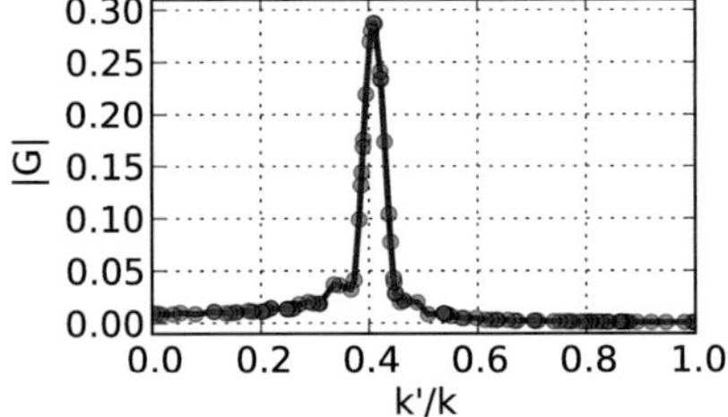

(b) HSP Geometry factor for two axially aligned TE$_{1,0,15}$ resonators.

Figure 2.5.: Rectangular waveguide resonators can be used for a LSW experiment, searching for HSPs. In the limit of a very high order mode, this configuration becomes identical to a laser LSW experiment.

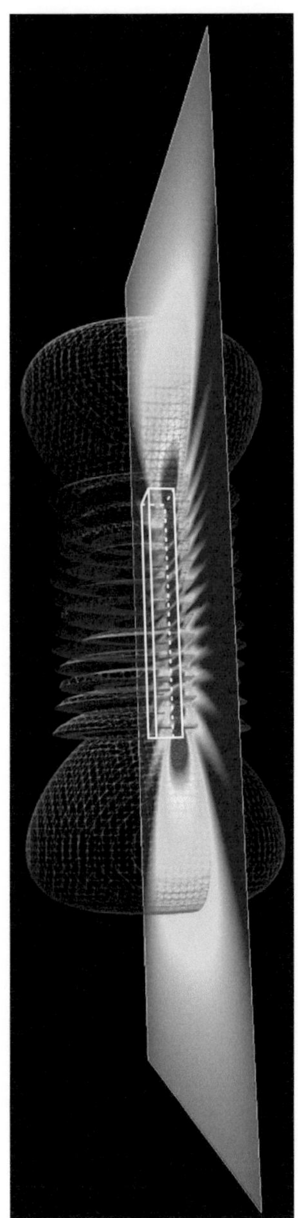

Figure 2.6.: Intensity of the predicted HSP field for the $TE_{1,0,15}$ mode of a rectangular waveguide resonator for $k'/k = 0.4$. Intensity is shown colour coded on a cutting plane and as a 3D iso-contour for a specific intensity value. Note how the field is more intense in axial direction with the waveguide.

3. Design and realization of the experimental setup

While the experiment has been described from a more theoretical point of view in the previous chapters, we will discuss here the engineering aspects and the practical side of designing a microwave based LSW experiment.

The four critical engineering challenges can be summarized as:

Resonant cavities: The microwave cavities are – figuratively speaking – the antennas for WISPs. They stimulate the conversion from photons to WISPs and back. Both cavities must operate with a well defined resonant frequency, quality factor and coupling to a certain mode. Details of the design process and their construction are given in 3.2.

EM shielding: The design and realization of the shielding enclosures has been described in 3.4. Their purpose is to reduce the RF field strength between the cavities by more than 300 dB, allowing us to achieve the intended sensitivity to WISPs while excluding any false observations due to EM leakage.

Microwave front-end: The electronic components immediately downstream of the detecting cavity are referred to as the microwave front-end. Its function is to amplify and transmit the hypothetical WISP signal over an optical fibre, while only adding a minimum amount of additional noise. Details are given in 3.5. For axion measurement runs, both cavities have been placed in a 3 T superconducting magnet – an environment demanding special requirements from the front-end components.

Signal receiver: The signal processing involves frequency conversion to baseband, digitization and narrowband filtering. These steps are explained in 4. We were able to detect sinusoidal signals with a power of less than -210 dBm which corresponds to 0.5 microwave photons per second for a 3 GHz signal.

3.1. Overview of the measurement setup

A detailed schematic of the complete experimental setup can be seen in Fig. 3.1. This configuration emerged – after numerous iterative improvements – as the final setup which was used for the most sensitive HSP and ALP measurement runs. The apparatus has been designed to be reasonably easy to transport and fast to set up, allowing us to use a superconducting magnet from an external partner for the ALP measurement runs.

A condensed overview of all the involved components, their arrangement in the experimental setup and their function shall be given here.

Outside the magnet, a commercial signal generator (ANAPICO APSIN6010) generates a sinusoidal RF signal of frequency f_{sys} ($\approx$ 1.7 GHz for ALP search), which is amplified to about 50 W of RF power by a solid-state power amplifier. A $\approx$ 3 m long coaxial cable transmits the signal to the emitting cavity (1) within the superconducting magnet. The transmission line has a corrugated outer conductor of 50 mm diameter to minimize RF losses. A directional coupler allows us to monitor the incident and reflected RF power between power amplifier and emitting cavity. The RF signal excites a resonant mode in the emitting cavity volume, resulting in a very strong electromagnetic field. The electric field of this mode is aligned with the external magnetic field from the superconducting magnet in a way to maximize the "production" of ALPs (see 3.2.2). During nominal operation, $\approx$ 90 % of the availible RF power is coupled into the cavity and subsequently absorbed by resistive losses in the cavity walls. Hence the emitting cavity will heat up to an operating temperatur of 50 °C - 90 °C, depending on air flow and temperature of the environment.

The detecting cavity is also placed in the magnet, in immediate vicinity (20 mm) to the emitting cavity. A hypothetical ALP – converting back to a photon – would excite a resonant mode at the frequency f_{sys}. If this happens, a very faint microwave signal can be coupled out from the detecting cavity. It is amplified by a low noise amplifier (3), filtered to

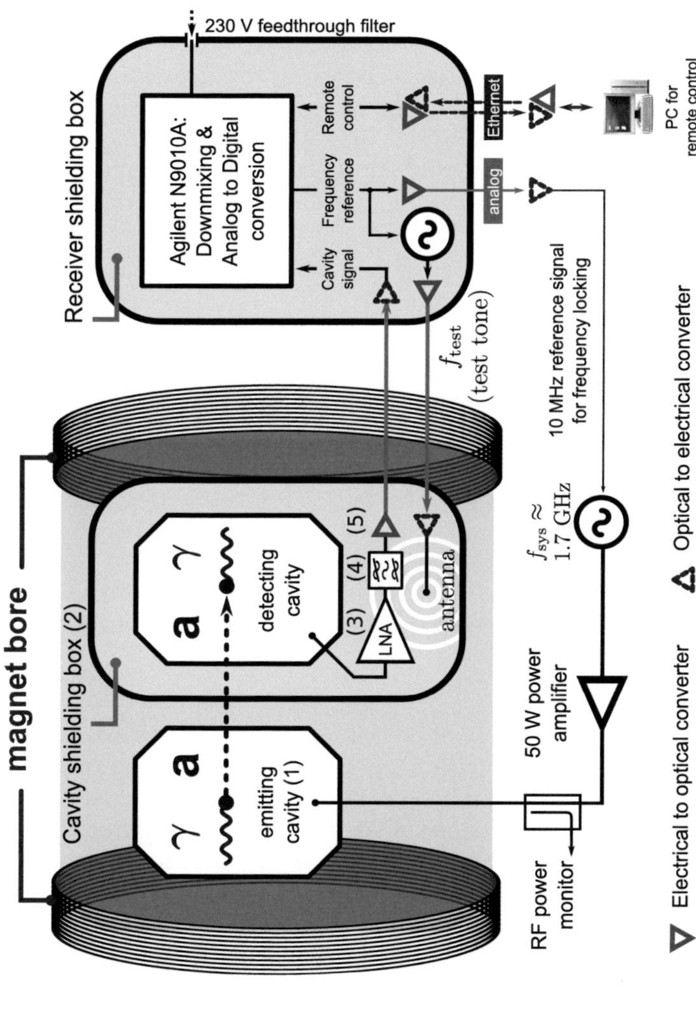

Figure 3.1.: Detailed block diagram of the microwave LSW experiment, as it has been carried out in June 2013 to search for ALPs.

reduce the background noise power (4) and then modulated on an optical carrier by an optical transmitter (5).

All components on the detecting side of the experiment are enclosed by an EM shielding enclosure (2). It prevents electromagnetic leakage from coupling into the detecting side and distorting the experimental results. Interference can originate from the emitting cavity or from ambient sources. The shielding enclosures can only achieve the required amount of EM attenuation if all analog and digital signals are transmitted over optical fibres. Commercial analog optical link modules have been used for that purpose. In addition, the optical transmitter and receiver within the magnet had to be modified to ensure their compatibility with the 3 T field.

The noise like signal from the detecting cavity is transmitted over an optical fibre to the "Receiver shielding box", which is placed outside of the magnet. It contains a Vector Signal Analyzer (VSA), which converts the signal to baseband, digitizes and records it. Remote control of the VSA within the shielded environment is made possible by a commercial ethernet-over-fibre transceiver. The acquisition memory of the VSA limits the maximum length of a measurement run to ≈30 h for recording with a nominal 2 kHz bandwidth. After the measurement run, the recorded data is processed offline, searching for sinusoidal signals in the background noise.

To test if the receiving chain is operational and the EM shielding is leak-tight, a test tone is emitted within the cavity shielding enclosure by an antenna. The signal couples through the shielding into the components of the RF front-end. The test tone is visible in the recorded physics data and hence indicates that the receiving chain was working. If the signal is not visible, a problem with the receiver is indicated. An increased test tone amplitude by several orders of magnitude would indicate a potential problem with the electromagnetic shielding (for example a loosely connected SMA connector). Furthermore, the VSA provides a 10 MHz clock signal, which is provided to all RF oscillators as a common frequency reference. This mitigates relative frequency drifts over time, which is an important requirement for the narrowband detection concept.

3.2. Microwave cavities

Microwave cavities are used to gain a resonant enhancement during photon to WISP and WISP to photon conversion on the emitting and detection side of the experiment. This can be understood from a particle point of view: Instead of propagating through the active region (the magnet for the ALP case) only once, the cavities capture the photons and keep them there for a short amount of time.

Our goal is to maximize the experiments sensitivity to ALPs or HSPs. The parameters of the two cavities influence this sensitivity in a significant way. From Eq. 2.15 and Eq. 2.16 we can deduce several requirements and important points, which we are taking into account when designing the cavities:

RF power: More RF power corresponds to coupling more photons into the emitting cavity, making it more likely to produce a WISP. RF power is limited by the thermal dissipation capability and the maximum tolerable electric field strength.

Quality factors: A larger Q factor corresponds to a longer photon lifetime within the cavity, making it more likely for a particular photon to convert to a WISP (or vice versa). Normal conducting cavities have been used, so the Q factors are limited by resistive losses in the wall material. Superconducting cavities have been discussed in 3.2.1.

Geometry: The geometric form factor (G) encodes the position of the cavities relative to each other, the electric field configuration within and the direction of the external magnetic field (only for ALPs). Generally, the distance between the cavities needs to be minimized. Different modes couple with different strengths to HSPs or ALPs. Some higher-order modes even show directionality. The G factor has been analyzed in 2.4.

Frequency: A lower system frequency (f_{sys}) results in a larger cavity – hence a larger detection volume. Furthermore, the lower quantum energy allows us to produce a larger number of photons for a given input power. Therefore LSW experiments are more sensitive at lower frequencies (see also Eq. 2.15). Nonetheless, the magnet must be large enough to accommodate the cavities and the lower quantum energy of the photon (E_{ph}) will restrain the maximum detectable WISP mass.

In 3.2.1 we start by investigating whether superconducting technology is relevant for WISP search and in particular for the CROWS experiment. A suitable cavity geometry is discussed and the mechanical design process is presented in 3.2.2. Design and construction of the additional parts like the RF power coupler and the tuning mechanism is shown in 3.2.3. The most sensitive modes for ALP or HSP search need to be identified, which is the topic of 3.2.4. Once a working mode has been found, its quality factor needs to be determined with high accuracy – our measurement process is explained in 3.2.5. The operational experience during the experimental run and a method for measuring the resonant frequencies "online" is detailed in 3.3.

3.2.1. Superconducting cavities

Superconducting cavities are nowadays frequently used in particle accelerators. Electric field gradients of > 25 MV/m and unloaded quality factors of $Q_0 = 10^{10}$ are routinely achieved, for example in TESLA accelerator cavities [73]. Although these cavities are not compatible with strong magnetic fields and therefore cannot be used for an ALP search, they could be the key ingredient for a very competitive HSP experiment. Note that the quality factor of emitting and receiving cavity are, together with the geometry factor, the most significant parameters affecting the experiments sensitivity to HSPs (as indicated by Eq. 2.16). In fact, superconducting cavities could increase sensitivity by 3 orders of magnitude compared to the CROWS experiment described in this work.

Nonetheless, a solution to a significant engineering challenge needs to be found first: the stabilization of the cavities resonant frequency. The experiments sensitivity will degrade if the resonant frequency of both cavities does not agree with the system frequency f_{sys} during the whole measurement run. Cavities with higher Q factors are more sensitive to drifts in resonant frequency (f) due to mechanical deformations (l), vibrations or temperature changes (T). The relation is given by:

$$\frac{1}{Q} = \frac{\Delta f}{f} = \frac{\Delta l}{l} = \frac{\Delta T}{T}.$$ (3.1)

Therefore the very high quality factors of superconducting cavities $O(10^{10})$ demands proportional requirements for their mechanical stability in the order of $O(10^{-10})$. It can be seen that even the smallest vibrations (e.g., through a nearby vacuum pump) can easily lead to detuning issues.

It is more feasible to use superconducting technology on the emitting side of the experiment. For example, a feedback driving scheme [74] could utilize the cavity itself as frequency determining element. However, the advantage of superconducting technology is not fully exploited by using it exclusively on the emitting side. Note that a stronger electromagnetic field within the emitting cavity – which is significant for HSP sensitivity – can also be achieved by more RF power instead of higher Q factors.

The real benefits of superconducting technology would be exploited on the detection side of the experiment, where the higher Q-factor directly enhances the conversion probability from HSPs to photons. The most significant engineering challenge would be to actively control the detecting cavities resonant frequency to mitigate frequency drifts. A closed loop feedback system might be used for that purpose, which consists of a measurement device and an actuator. The latter might be implemented with the required precision by a tuning mechanism which elastically deforms the cavity with piezo actuators. The more challenging aspect is to measure the resonant frequency of the relevant mode fast and precisely enough to allow realtime control of its resonant frequency. This is problematic due to the fact, that the detecting cavity must be free of signals in the frequency range where we would expect a HSP signal. For example, it is not possible to determine the resonant frequency in a VNA measurement, as the incident RF test signal would interfere with HSP detection.

Note that accurate control of field strength and phase in particle accelerator RF cavities is usually achieved by overcoupling them to the power amplifiers. This leads to a broader bandwidth and reduced loaded quality factor operation. For example, the nominal loaded quality factor of the superconducting LHC cavities is $Q_L \approx 20000$ [75].

For the above reasons, the decision was made to use normal conducting cavities in the first design iteration of the CROWS experiment. In future experiments, the advantages of superconducting technology might be exploited.

3.2.2. Geometry

The starting point for the cavity design is a cylindrical geometry, operating in the 1-3 GHz range. This choice keeps the setup on a table-top scale and ensures that the inset will fit in the bore of most MRI magnets.

While every cavity resonator has an infinite number of resonating modes at different frequencies, we are required to operate on only a single one of them, which must be clearly distinguished in frequency to the neighboring ones. Solving Maxwell's equations for a cylindrical geometry, an analytical expression can be derived for the resonant frequency of an arbitrary mode [76]:

$$f_{nml} = \frac{c}{2\pi\sqrt{\mu_r\epsilon_r}}\sqrt{\left(\frac{p_{nm}}{r}\right)^2 + \left(\frac{l\pi}{h}\right)^2}, \qquad (3.2)$$

where c is the speed of light, μ_r and ϵ_r are the relative permeability and permittivity constants of the homogeneous filling material, r is the radius and h is the height of the cavity. The integer mode numbers n, m and l define the mode index in a cylindrical coordinate system. Note that there are two sorts of resonating modes, the ones with a purely transverse electric field (TE) and the ones with a purely transverse magnetic field (TM) [76]. For TM modes, p_{nm} is the m-th root of the Bessel function of the first kind. For TE modes, p_{nm} is the m-th root of the derivative of the Bessel function of the first kind. From Eq. 3.2 we can derive two observations. Firstly, it can be seen that the resonant frequency scales proportionally to the linear dimensions of the cavity. While a lower operating frequency provides an advantage in sensitivity (see 2.3), the physical size imposes a limit to this approach. Secondly, higher order modes are more tightly packed in frequency. This sets an upper limit on operating frequency, as modes of higher order are more difficult to excite separately in a well defined way.

The cavities quality factor (Q) is defined as:

$$Q = 2\pi f \frac{\text{mean}\,(\mathbf{W})}{\mathbf{P}},$$

(3.3)

where $\mathbf{W}$ is the mean energy stored in the electric and magnetic field and $\mathbf{P}$ is the loss of power [77]. The latter is dominated by resistive losses of the displacement currents in the cavity walls. To obtain high Q factors, we need to maximize the volume (related to $\mathbf{W}$) while keeping the surface area (related to $\mathbf{P}$) low. Hence the optimum shape would be a spherical resonator – which is difficult to manufacture accurately. A cylindrical shape has been adopted for the cavities as it represents a good trade-off with a small surface area while still being easy to manufacture. To approximate the spherical shape, the aspect ratio $2a/h$ has been set to ≈ 1.

The TE_{011} mode provides a comparatively high quality factor because of its low Ohmic losses and the favourable distribution of its displacement currents. They flow along the circumference of the cylinder and do not have an axial component – hence they do not cross the boundary between the two half shells, which will make up the cavity. This provides a big advantage, making losses due to the contact resistance between the half-shells completely negligible. For that reason, the TE_{011} mode is widely used in microwave absorption wavemeters, in microwave filters and is also a very good candidate for WISP search. However, in a purely cylindrical geometry, this mode is degenerate with the TM_{111}, meaning they share the same resonant frequency and hence exchange energy. To make the mode accessible for WISP search, the geometry of the cavity has been modified. Its top and bottom edges have been strongly chamfered. From a mechanical perspective, this provides a transitional shape between a cylinder and a sphere. While the TE_{011} mode is degenerate with a different mode for the cylindrical and the spherical case, the degeneracy does not appear for the transitional shape [78]. A beneficial side effect of the spherical-like geometry is the slightly increased quality factor.

The final geometry was found in several iterations, making use of a commercial full-wave simulation software. The inside dimensions are given in Fig. 3.2. For manufacturing, the cavities were divided into two half shells

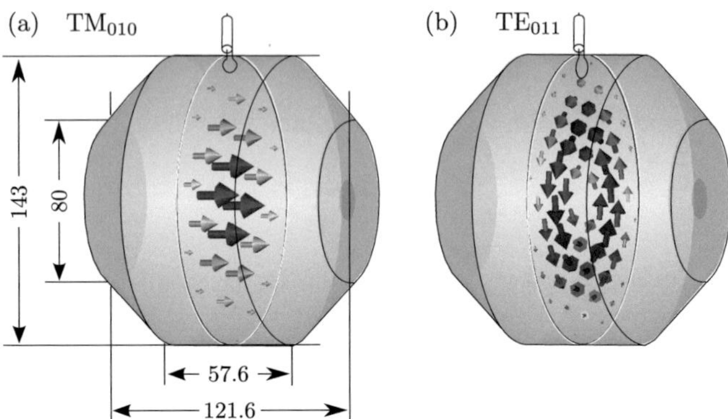

Figure 3.2.: Inside dimensions of the cavity in [mm]. The coupling loop can be seen on top. The electric field configuration of two modes is shown on a transverse cutting plane. (a) TM$_{010}$ mode for ALP search, (b) TE$_{011}$ mode for HSP search.

and each was machined at CERN from brass material. A photo of the machined pieces can be seen in Fig. 3.3(a).

The surface properties of the inner cavity walls have a significant impact on the quality factor. By applying a 10 μm thin silver coating, a $\approx$ 2-fold improvement in Q was achieved. Due to the skin effect, most of the displacement currents of the mode flow very close to the surface and therefore in the silver layer with comparatively low resistive losses. The skin depth δ_s gives an indication of how deep the currents penetrate the cavity walls:

$$\delta_s = \sqrt{\frac{1}{\pi f_{\mathrm{sys}}\mu\sigma}} \approx 1.5 \ \mu\mathrm{m}, \quad I(d) = I_0 \cdot (1 - e^{-\delta_s/d}), \qquad (3.4)$$

where $\sigma = 6.3 \cdot 10^7$ S/m is the conductivity of silver. The equation predicts, that the current density I at the boundary between the brass and silver layer (d= 10 μm) has decayed to less than 14 % of its initial value.

When exposed to the atmosphere, silver tends to form an oxide layer, which is disadvantageous for its electrical properties. To prevent this oxidation,

Figure 3.3.: (a) The cavities after machining from brass material. (b) Wire-loop antenna for critical coupling to the TE_{011} or TM_{010} mode. (c) Open half shell after surface treatment. The coupling loop and the tuning screw are visible.

a $\ll 1$ μm thin flash of gold has been deposited on top of the silver layer by electroplating.

3.2.3. Coupler and tuner

To couple power in or out of the cavity in an efficient way, the electromagnetic fields need to be matched to the characteristic impedance of a 50 Ω coaxial transmission line. This is achieved by a small antenna, consisting of a simple wire loop of ≈ 8 mm diameter. The couplers were constructed by soldering a semi-rigid coaxial cable in an axially drilled M8 brass screw as shown in Fig. 3.3(b). The inner conductor was bent into a wire loop on one end and a SMA connector was attached on the other end. The couplers are mounted in a corresponding threaded hole on the cavities. The coupling strength (β) is determined by the flux of the magnetic field of a specific mode through the wire loop. Adjustment of the insertion depth or the angle of the screw changes the effective surface area of the loop and hence the coupling strength. Critical coupling ($\beta = 1$) was accomplished by tuning the couplers with a Vector Network Analyzer (VNA) for minimum reflection – typically more than 30 dB of reflection loss

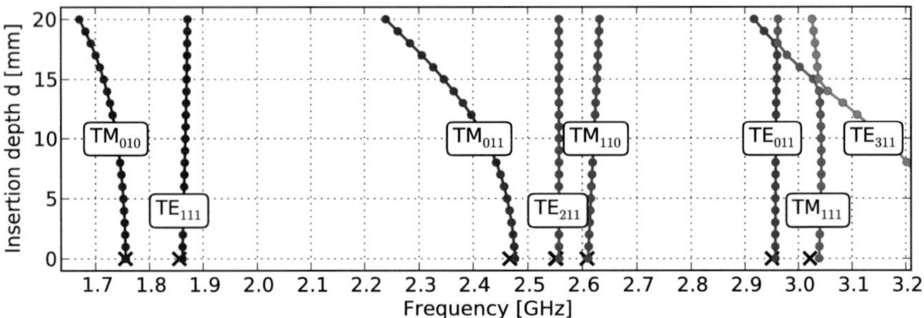

Figure 3.4.: Measured cavity mode chart as a function of tuning screw position d. For d=0 mm, black crosses indicate corresponding simulation results.

has been achieved. After tuning, the couplers were mechanically clamped in place with locking nuts.

To compensate manufacturing tolerances or temperature drifts, it is necessary to adjust the cavities resonant frequency in a narrow range. Therefore a tuning screw of 10 mm diameter has been mounted axially on one of the half shells. The fine threaded screw penetrates the cavity volume and slightly perturbs the fields within. Depending on the nature of the perturbed field (magnetic or electric), the frequency of a certain mode can be tuned up or down. For the TE$_{011}$ mode, this provides us a tuning range of $\approx$ 10 MHz, which turned out to be sufficient for all operating conditions. A measurement of the frequency dependence of the first seven modes as a function of the tuning screw position is shown in Fig. 3.4. The figure also indicates that there are no mode crossings within the utilized tuning range from 0-10 mm.

3.2.4. Choice of mode

A particular cavity mode needs to be chosen for the experimental measurement run. This choice is important, as the mode significantly influences the detection sensitivity in two ways: by its quality factor (Q) and its geometric form factor (G). The squared magnitude of both parameters

Table 3.1.: Comparison of cavity modes for ALP and HSP search by a Figure Of Merit (FOM). The cavity dimensions are identical to the one in Fig. 3.2. Unless otherwise noted, the cavities have been aligned axially with a distance of 20 mm between their inner walls.

Mode	$Q_0/2$	$|G|_{\mathrm{ALP}}$	$|G|_{\mathrm{HSP}}$	FOM_{ALP}	FOM_{HSP}
TM_{010}	11 800	0.97	0.79	**11 500**	9 300
TE_{111}	12 700	0.01	0.88	100	11 200
TM_{011}	8 100	0.58	0.21	4 700	1 700
TE_{211}	12 800	0.00	0.45	0	5 800
TM_{110}	11 400	0.20	0.76	2 300	8 700
TE_{011}[1]	23 200	0.00	0.52	0	**12 100**
TM_{111}	10 800	0.06	0.14	324	800

affects the sensitivity as indicated by Eq. 2.15 and Eq. 2.16. To find a good choice for ALP and HSP search, we have derived the Q and G parameters for several mode-candidates and evaluated the product $Q \cdot |G|$ as a figure of merit (FOM) of their suitability. The unloaded quality factor (Q_0) has been obtained by a transmission measurement with a VNA, using two weakly coupled antennas in the cavity. The antennas are ≈ 2 mm short wire stubs, which influence the modes in the cavity only negligibly due to their very weak coupling. Hence the result in the Table is shown as $Q_0/2$, which corresponds to the loaded quality factor of the critically coupled cavity. The geometric form factor has been calculated for several modes by numerically evaluating the relevant overlap integral – details to the procedure are given in 2.4. Unless otherwise noted, we have assumed that the cavities are aligned axially (like shown in Fig. 3.2), with 20 mm distance between them. For $|G|_{\mathrm{ALP}}$, the external magnetic field has been assumed to point in the axial direction. As $|G|$ is a function of WISP mass, only its maximum value has been utilized for calculating the FOM. The results for each mode are shown in Table 3.1. The Table suggests that among the analyzed modes, the TM_{010} is favourable for ALP search and the TE_{011} for HSP search.

---

[1] Cavities placed next to each other with 20 mm distance.

This is plausible and can be understood intuitively: For ALPs we achieve good geometric factors if the electric field in the cavity is parallel to the external static magnetic field. Only the fundamental TM_{010} mode provides an E-field which points everywhere in the same direction and can therefore be aligned over the largest possible volume. For HSP search, the E-field does not need to be aligned and Table 3.1 shows, that all of the geometry factors are relatively large. This is expected as the modes are of low order and cannot radiate HSPs with significant directionality (see also 2.4). Nonetheless, there is a clear advantage in using the TE_{011} mode because of its significantly larger Q, which overcompensates its slightly lower geometric factor. The surface currents of this mode flow along the circumference of the cavity and do not cross the boundary between the two half shells [79], which reduces resistive losses as compared to the other modes.

3.2.5. Measurement of quality factor

The loaded quality factors significantly influence the sensitivity of the experiment and therefore had to be determined with high accuracy. The frequency dependent reflection coefficient Γ has been measured with a Vector Network Analyzer (VNA). The cavity parameters Q_0, Q_L, β and their respective uncertainties have been extracted from the VNA data by means of a curve fitting algorithm, described in [80]. The resulting quality factors for the ALP and HSP measurement runs can be found in Table 5.2 and Table 5.3. A typical result, taken immediately before the ALPs run in June 2013 is shown in Fig. 3.5.

The coupling loops in both cavities are made from copper wire – hence we expect a small amount of signal power to be lost due to its finite conductivity. To estimate the amount of coupling loss, we evaluate the reflection coefficient far-off resonance. In the ideal case it would be equal to $|\Gamma| = 0$ dB (a short or open circuit). We measured a value of $|\Gamma| = -0.1$ dB. This is also visualized in Fig. 3.5 in the Smith chart: due to the coupling loss, the cavity circle does not quite touch the outer circumference of the chart. As the amount of coupling loss is negligibly small and as it is already implicitly included in the loaded quality factor (Q_L) for the

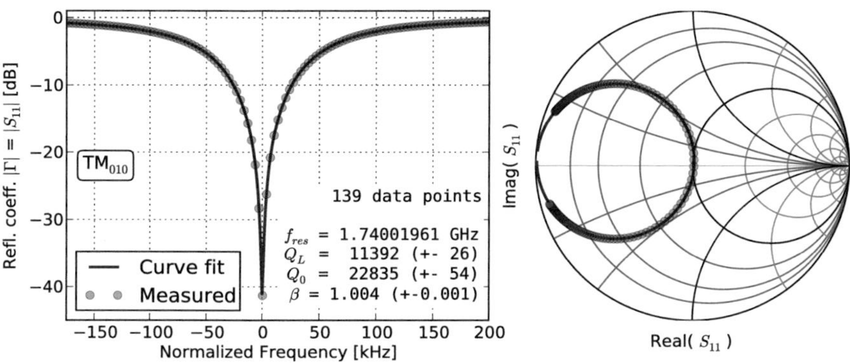

Figure 3.5.: Result of a $S_{11} = \Gamma$ measurement of the detecting cavity, tuned to the TM_{010} mode. The cavity parameters have been extracted by means of a curve fitting procedure, adapted from [80]. Left: magnitude of $|\Gamma|$ indicates > 30 dB return loss. Right: the complex Γ on a Smith chart.

critically coupled cavity, we do not have to consider coupling loss separately for the detection sensitivity of the experiment.

Note that both cavities provide $|\Gamma| < -30$ dB return loss on their resonant frequencies. Therefore signal losses through impedance mismatch can be neglected under the assumption that the resonant frequency agrees with the excitation frequency f_{sys}.

3.3. Cavity operation and monitoring

WISPs could exist within an enormous parameter space, from which only a small subset is probed by the CROWS experiment. The outcome of an experimental run is therefore more likely to be an exclusion result than a discovery. From an experimental perspective, producing an exclusion result is more challenging than detecting a WISP candidate. In the latter case, the WISP signal would be visible in the spectrum and would implicitly confirm, that the setup is working properly. On the other hand, in the case of an exclusion result, there would not be an obvious indication if there is really no WISP or if a fault condition in the apparatus was suppressing the detection of a potential candidate. To prevent this and to prove that the setup was actually operational, several of the critical parameters were recorded during the entire duration of a measurement run. The resonant frequency (f_{res}) of both cavities is an example of such a critical parameter: if one of them drifts off-tune and its resonant frequency does not agree with the system frequency (f_{sys}), the sensitivity of the experiment will degrade. This could disguise a potential WISP candidate and lead to an invalid exclusion result. Therefore, for the calculation of an exclusion limit, the maximum deviation of f_{res} needs to be measured, recorded and taken into account.

The resonant frequency of the cavities has been estimated from three different observables:

RF power: The emitting cavity has been monitored by logging the incident (P_{inc}) and reflected RF power (P_{refl}) at the cavities coupling port.

Noise power: The detecting cavity has been monitored by evaluating the spectral background noise in the recorded data. As it is dominated by the thermal noise emitted from the cavity, it allows us to determine its resonant frequency.

Physical temperature: Both cavities have been monitored by measuring their physical temperature. The change in resonant frequency is directly proportional to the change in temperature, allowing us to state the maximum deviation of the resonant frequency during the measurement run.

3.3.1. Emitting cavity

The RF power amplifier excites a microwave signal at the frequency f_{sys}, which propagates over a coaxial cable to the emitting cavities coupling antenna. Part of this incident wave is coupled into the cavity and subsequently absorbed by the walls – another part is reflected back towards the amplifier. The ratio of the reflected to the incident power ($P_{\mathrm{refl}}/P_{\mathrm{inc}}$) is given by the reflection coefficient $|\Gamma|^2$, which depends on the cavities

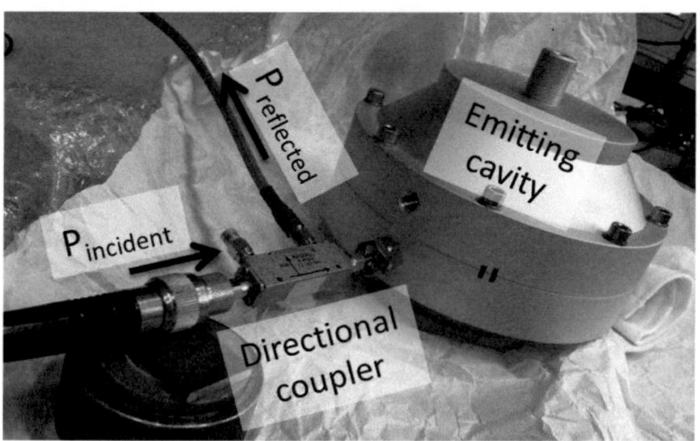

Figure 3.6.: The resonant frequency of the emitting cavity is monitored by means of its reflected RF power, which is measured with a directional coupler.

resonant frequency (f_res) in relation to the excitation frequency (f_sys), the loaded quality factor (Q_L) and the coupling coefficient (β) by:

$$\frac{P_\mathrm{refl}}{P_\mathrm{inc}} = |\Gamma(f)|^2 \approx \left| \frac{2\beta}{\beta + j2Q_L f_\mathrm{norm} + 1} - 1 \right|^2, \qquad (3.5)$$

$$f_\mathrm{norm} = \frac{f_\mathrm{sys} - f_\mathrm{res}}{f_\mathrm{res}}.$$

The derivation of this equation has been shown in A. Measuring and recording the incident and reflected power at the cavities coupling port allows us to make a statement about the tuning (f_norm) of the cavity. Reflected power will only be minimum if $f_\mathrm{sys} = f_\mathrm{res}$. Detuning causes an increase in P_refl. If completely off-tune, all of the incident RF power would be reflected [77].

P_inc and P_refl have been measured with a dual-directional coupler, placed on the coaxial line between power amplifier and cavity as shown in Fig. 3.6. It provides a scaled-down version of the incident and reflected RF wave. The scaling factor corresponds to the coupling coefficients, which have been measured with a VNA and are shown in Fig. 3.7. Two crystal detector diodes (HP423A) convert the RF-signals from the directional

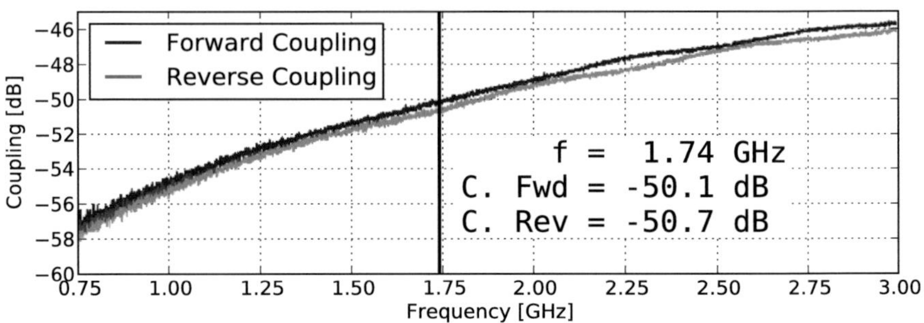

Figure 3.7.: Calibration factors for the directional coupler (ALPs, June 2013).

coupler to DC voltages, which were recorded by a data logging device (Picolog ADC-20). The non-linear response curve of the detector diodes has been measured beforehand and used as a calibration table, to linearize the diodes. This setup allows us to transform the measured voltages into incident and reflected RF powers on the cavities coupling port.

We define the absorbed RF power as $P_{\mathrm{EM}} = P_{\mathrm{inc}} - P_{\mathrm{refl}}$, which will be on the order of 50 W, when the cavity is operating on tune, driven by the nominal RF power and close to zero when the cavities resonant frequency is far off f_{sys}. An example of P_{EM} over time, as it was recorded during the ALPs run in June 2013 is shown as the green trace in Fig. 3.8. The average of P_{EM} has been utilized to determine the experiments sensitivity to WISPs.

During nominal operation, the emitting cavity dissipates $P_{EM} \approx 50$ W of heat by forced air cooling and reaches a steady state temperature of $\approx 50\,°\mathrm{C}$. No external temperature stabilization is used – instead we rely on a natural feedback effect to keep the cavity on tune. This effect can be observed once the cavity is warmed up and has reached thermal equilibrium. Then an increase in cavity temperature manifest itself in a lower resonant frequency (see Fig. 3.10), which in turn leads to less RF power being absorbed by the cavity (see Eq. 3.5). The temperature of the cavity decreases and the feedback loop is stable. Stability can only be achieved when the excitation signal falls on the upper half of the resonance curve.

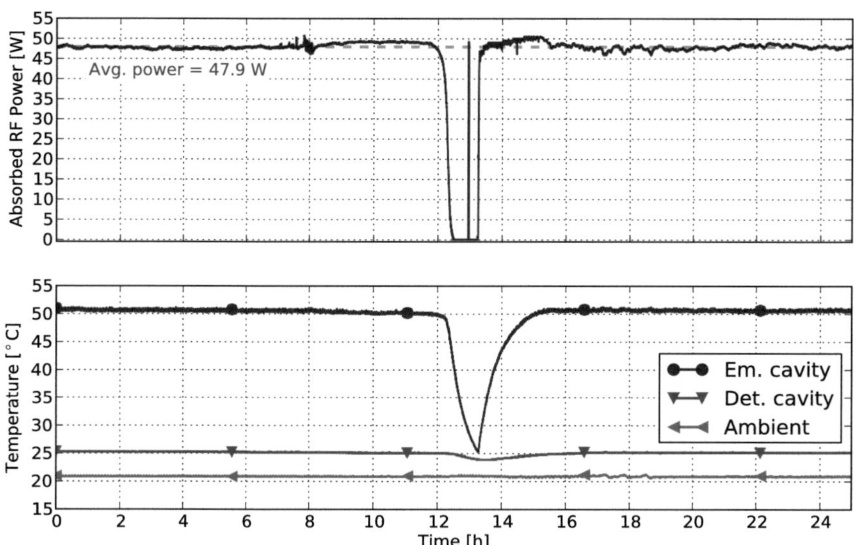

Figure 3.8.: Example record of the emitting cavities temperature and absorbed RF power during the ALP measurement run in June 2013. The emitting cavities resonant frequency exceeded the frequency of the excitation signal at t=12 h. This lead to a thermal runaway condition during which the cavity completely stopped to absorb RF power within minutes. The cavity subsequently cooled down, which also influenced the temperature of the detecting cavity due to its close vicinity.

To stay within that region, even with changes in ambient temperature, f_{sys} has been set slightly higher than f_{res}, leading to ≈ 5 W of constantly reflected RF power. The losses due to mismatch of the cavity and due to the attenuation of the coaxial cables have been taken into account for the exclusion limit calculation.

Exploiting the natural feedback effect is the most simple and effective way to operate the emitting cavity, nonetheless it suffers from one potential pitfall. The operating point of the cavity fluctuates slightly due to small changes in ambient temperature and input RF power. If the cavity cools down too much and its resonant frequency exceeds the frequency of the excitation signal (f_{sys}), the cavity will operate on the lower half of the resonance curve, where the slope is of opposite sign. This will lead to positive feedback operation. The cavity will enter a thermal runaway condition, detune, stop to absorb RF power and cool down to ambient temperature within ≈ 30 minutes. An example can be seen in Fig. 3.8, where thermal runaway happened after 12 h of stable operation during the ALPs run in June 2013. More details on how the cavity has been brought back to operating temperature and how the incident has been treated in the data evaluation are given in 5.1.

3.3.2. Detecting cavity

Unlike the emitting cavity, the detecting cavity is entirely passive and not driven by any external microwave source – this would interfere with WISP detection. Nonetheless, the cavity is at room temperature and therefore emits thermal noise, which provides a way to monitor its resonant frequency. Although the spectral noise power density of thermal noise is flat and frequency independent, the reflection coefficient of the cavity does make the noise frequency dependent. This has been derived in B. The noise power density we expect at the cavities coupling port (N_0) is:

$$N_0 = kT_{\mathrm{cav}}(1 - |\Gamma(f)|^2) \tag{3.6}$$

During the experimental run, N_0 has been measured. This required comparatively little effort, as the VSA is already connected to the receiving cavity for the purpose of recording experimental data and no changes to the hardware were necessary. A dedicated spectral noise measurement with a span of $SP = 1$ MHz has been carried out before and after each experimental run. In these traces, the complete resonant response is visible and a good estimate of the absolute resonant frequency can be determined from the point where the noise power is maximum. Note that this procedure assumes, that the physical temperature of the cavity walls is significantly higher than the noise temperature of the receiving chain. In our case, the former is 298 K and the latter 43 K, hence the assumption is met. A typical N_0 trace is shown in Fig. 5.10, which clearly indicates the resonant frequency of the detecting cavity.

Unfortunately, the VSA cannot carry out a dedicated spectral noise measurement ($SP = 1$ MHz) and record physics data ($SP = 2$ kHz) at the same time. We are hence restricted in measuring the state of the cavity to in-between recordings. To get a indication of the state, also during the measurement run, we evaluate the recorded physics data. This provides us a time-trace of the average noise power in a $SP = 2$ kHz band around f_{sys}. The recorded span is not wide enough to resolve the resonance peak of the cavity (with a 3 dB bandwidth of ≈ 75 kHz) on a frequency scale. Nonetheless, the average noise power will be at a maximum if the cavity is on resonance and decrease by > 10 dB, if the cavity is severely detuned with respect to f_{sys}. A typical result is shown in Fig. 5.9(c).

Both, the frequency and time domain based results have been crosschecked with each other during the evaluation of each measurement run and found to agree within ± 0.25 dB. Therefore, monitoring the thermal noise power provides a precise tool to estimate the detecting cavities tune during a measurement run.

3.3.3. Thermal monitoring

The underlying reason for drift in the cavities resonant frequency is their sensitivity to temperature variations due to the thermal expansion and contraction of their wall material. The relationship between resonant frequency and temperature is a highly linear one. Therefore, if the proportionality constant would be known, we could estimate the state of the cavities from a temperature measurement.

We have measured the sensitivity to temperature changes by heating one of the cavities to about 45 °C and then letting it slowly cool down to ambient temperature. The setup is shown in Fig. 3.9. During the cooldown phase, its temperature and resonant frequency were measured. A VNA was utilized to measure S_{11}, from which the resonant frequency has been deduced in a similar fashion as described in 3.2.5. The parametric result, plotted as f_{res} over T is shown in Fig. 3.10. The slope of the trace relates changes in temperature to changes in resonant frequency.

During the WISP measurement runs, the physical temperature of both cavities has been measured with high precision by LM35 sensors (HSP

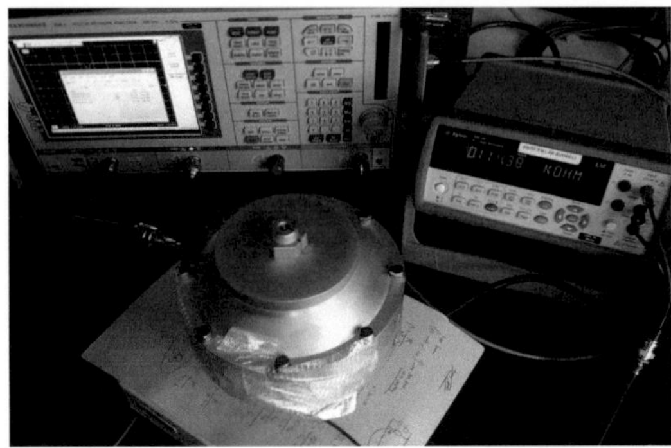

Figure 3.9.: Measurement setup for determining the cavities sensitivity constant, relating changes in temperature to changes in resonant frequency. The cavity has been heated, after which it slowly cooled down to ambient temperature. Its resonant frequency has been monitored by a VNA and its temperature by a LM35 sensor during the cool-down phase.

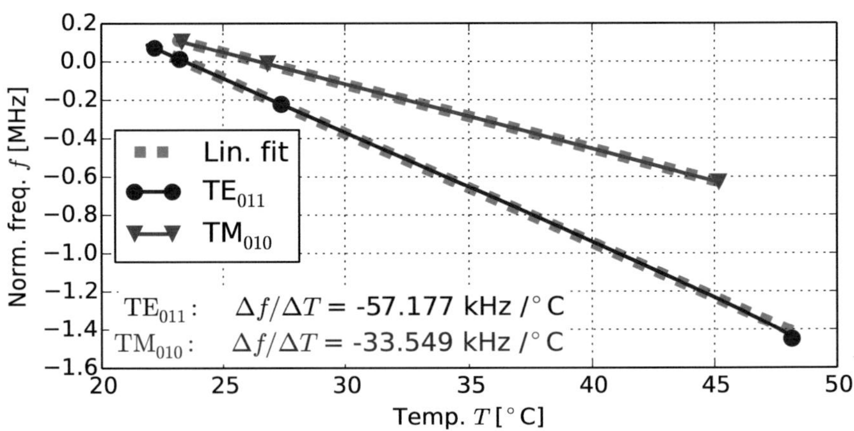

Figure 3.10.: Resulting calibration factors, relating temperature change to frequency change.

runs) or by an optical fibre based thermometer (ALP runs). While the sensors were placed in direct contact with the emitting cavity, this was not possible for the detecting cavity, as the cables would have to penetrate the EM shielding enclosure. Therefore the temperature was measured on the outside of the enclosure, which is in good thermal contact with the detecting cavity.

The estimated resonant frequency has been compared with the other methods. We found that the accuracy is good in general, however during steep temperature transients there is a discrepancy between the actual temperature of the cavity and the measured one, leading to an overestimation of the cavity drift. Nonetheless, stable temperature conditions indicate a stable resonant frequency.

It shall be noted as a concluding remark, that one way to significantly simplify the procedure of a measurement run and to make it more reliable would be to add a control system to actively stabilize the temperature of the two cavities. This would allow to eliminate any drift in resonant frequency. While in the current design a way was found to keep f_{res} stable in an entirely passive way, the system is very sensitive to changes in ambient temperature, which can lead to a thermal runaway condition as demonstrated by the ALP measurement run in June 2013. Furthermore, in the current design, the cavities need several hours before each measurement run to reach thermal equilibrium. To solve these problems, a small heating element, driven in a control loop, would allow to keep the cavities at a fixed temperature, several degrees above ambient. This upgrade would shorten the set-up time and improve the reliability of the apparatus by a large degree, while only requiring a minimum amount of engineering effort. Due to time constraints, this feature could not be implemented before the most sensitive measurement runs were carried out.

3.4. Electromagnetic shielding

For a LSW experiment in the microwave range, electromagnetic shielding is one of the primary concerns. In fact, previous microwave LSW experiments [63, 64] were limited in sensitivity by leaked signals, which could have been mitigated by appropriate EM shielding. A very important aspect

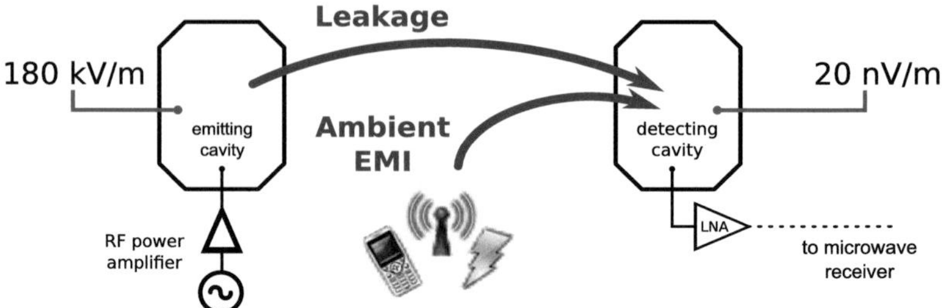

Figure 3.11.: Illustrating the two significant sources of EMI.

of the CROWS experiment was to design, build and measure a shielding concept which would allow us to avoid a potential bottleneck, limiting WISP detection sensitivity.

3.4.1. Theoretical considerations

In general, signals of longer wavelength are more difficult to shield. This is related to the effect of wave diffraction on obstacles. Consider the laser beam of an optical LSW experiment, it can be blocked completely by a simple metal plate. On the other hand, a microwave beam with a wavelength in the order of the metal plate would diffract at its edges and reach the space behind with barely any attenuation. Therefore diffraction is the primary reason why a comparatively complex 3 dimensional shielding enclosure is necessary for microwave LSW experiments.

As illustrated in Fig. 3.11, two sources of Electromagnetic Interference (EMI) need to be considered:

Ambient EMI: We assume that the experiment will be carried out in a laboratory environment, in absence of any provisions for EMI shielding against outside interferes. Ambient sources like cellphones, cellphone base stations, wireless network transceivers, radio and TV stations or airport radars will cause a certain level of background field strength, which potentially could interfere with the sensitive microwave detector and show up as random artifacts in the measured spectrum. For example, a mobile phone can transmit with up to

2 W output power into an omnidirectional antenna, resulting in a field strength of ≈ 3 V/m within a distance of 3 m.

Leakage: In the LSW experiment, emitting and detecting cavity have to be placed as close as possible to each other. A certain amount of electromagnetic leakage from the two cavities is unavoidable, which results in a weak coupling between them. Shielding is necessary to attenuate this coupling below the detection threshold of the microwave receiver. Leaked photons would show up at the same frequency and with the same shape in the spectrum as a potential WISP signal. It is not straightforward to discriminate between WISPs and leakage and thus shielding is critical to avoid false positive results in the experiment.

Shielding efforts have been concentrated on the detection side of the experiment, namely on the detecting cavity, the microwave front-end and the microwave receiver. This mitigates both, leakage and ambient EMI. No additional shielding around the emitting part of the experiment was applied for two reasons. Firstly, this would only help to reduce leakage and not ambient EMI. Secondly, it was measured that the field strength in vicinity to the emitting cavity and the power amplifier was low enough to be neglected with regards to the ambient EMI, even at full power operation.

From the expected field strengths in the emitting (180 kV/m) and detecting cavity (20 nV/m), we can estimate the amount of shielding required. Within 20 mm separation between the two cavities, we need a reduction of at least a $10^{13} = 260$ dB to produce meaningful results. Most microwave components used in the setup like SMA connectors or semi-rigid coaxial cables provide a screening attenuation of less than 120 dB, making an external shielding enclosure and strategic use of optical fibres for signal transmission necessary.

We define the shielding effectiveness (SE) as the relative reduction of electric or magnetic field strength in a certain point after applying the shielding:

$$SE = 20 \cdot \log \frac{E_0}{E_1}, \tag{3.7}$$

where E_0 and E_1 are the electric field strengths with and without shielding [81].

In practice, the electric or magnetic field strength in the near field can be measured with a calibrated probe antenna, optimized for this purpose. Note that electric and magnetic field strengths are related by the wave impedance Z_0. In the far field region (in a distance of $\approx 10\lambda$ from an emitter) the wave impedance is $Z_0 = \sqrt{\mu/\epsilon} \approx 377\ \Omega$ and the shielding effectiveness can be measured with either the magnetic or electric field, giving the same result.

The shielding needs to be effective for microwave signals in the GHz range. In this regime, shielding attenuates signals by two mechanisms. A large part of an incident plane wave will be reflected, because the metallic walls of the shielding enclosure provide a large step in surface impedance at their boundary to air. This mechanism is more effective to shield far fields or near fields which are dominated by their electric field component, having a wave impedance of $\geq 377\Omega$. Furthermore, the incident plane wave will be partly absorbed because of induced surface currents in the material and their ohmic losses. The absorption mechanism is more effective against near fields, dominated by their magnetic field component, having a smaller wave impedance than $377\ \Omega$.

The skin depth δ_s in the material is of significance, describing the depth in the material where the surface currents have fallen off to $1/e \approx 37\ \% \approx -9$ dB of their original value. It is given by

$$\delta_s = \frac{1}{\sqrt{\pi f \mu \sigma}}. \tag{3.8}$$

(see also Eq. 3.4) and in the order of a few μm for a metal of good conductivity like aluminium at 3 GHz.

In general, metals with high conductivity and high permeability provide better shielding due to more reflection and less skin depth (current flows closer to the surface). Although metals with large permeability (e.g., mu-metal, $\mu_r \approx 20000$) are very useful for shielding DC and slowly varying magnetic fields, they exhibit $\mu_r = 1$ at microwave frequencies and are thus inferior to copper or aluminium, because of their lower conductivity.

In conclusion, the choice of material for the shielding enclosure is not very critical at frequencies in the GHz range. The material should have a high conductivity and a thickness of $\geq 20\sigma$. We have chosen aluminium with a wall thickness of 5 mm, fulfilling these requirements and additionally providing good mechanical properties.

An important observation is that the weakest point of an electromagnetic shielding enclosure will determine its overall performance. For a removable enclosure, which is not hermetically sealed and not permanently welded together, the weakest points will be leakage through seams, gaskets, feedthrough ports and other openings. The impact of any of these components on the shielding effectiveness of the whole system (SE_{sys}) is expressed by:

$$\frac{1}{SE_{\text{sys}}} = \frac{1}{SE_{\text{wall}}} + \frac{1}{SE_{\text{aperture}}} + \frac{1}{SE_{\text{gasket}}} + \frac{1}{SE_{\text{filter}}}. \tag{3.9}$$

The components with the biggest impact on performance are:

SE_{wall}: A poor choice in material (too thin or too low conductivity) can lead to EMI leaking through the bulk wall material of the shielding enclosure. This is generally not a problem at GHz frequencies, where skin depths are usually in the order of a few μm.

SE_{aperture}: Usually openings need to be cut in the enclosure for ventilation, as optical windows or for mounting feedthrough filters and other components. These apertures will naturally have a lower shielding effectiveness than the wall material. Furthermore, unidentified gaps in the shielding or metal joints with bad contact are unwanted apertures. This kind of leakage can be effectively localized with a near field probe antenna.

SE_{gasket}: As the enclosure needs to be be removable, gaskets are essential to provide a continuous conduction path for currents across the lid to the mating surface. Contact springs or knitted wire meshes can be used. Surface corrosion and mechanical fatigue can cause bad contacts over time, weakening the shielding effectiveness of the gasket.

SE_{filter}: Feedthrough filters need to be used to provide the components in the enclosure with power. These are usually lowpass filters, conducting DC but blocking high frequency EMI, which would otherwise propagate over the copper cables into the shielded domain.

3.4.2. Construction of the receiver shielding box

The EM shielding has been split into two separate enclosures as the receiving electronics and the VSA cannot be placed in the strong magnetic field. The receiver shielding box, shown in Fig. 3.12a, is placed outside of the magnet and contains the instrumentation needed to detect weak microwave signals. The cavity shielding box, shown in Fig. 3.12b, is placed in a strong magnet during ALP measurements and contains the detection cavity including the RF front-end. The RF signals between the two shielding boxes are transmitted by optical fibres utilizing analog transceivers.

(a) Receiver shielding box   (b) Cavity shielding box

Figure 3.12.: Two shielding enclosures are necessary. (a) is placed outside the magnet and contains the microwave receiver. (b) is placed within the magnet and contains the detecting cavity.

The receiver shielding box needs to be rugged and easy to transport with the instruments inside. It is also required to support the airflow from an external fan, as the instruments within dissipate several hundred watts of heat.

It was constructed from a straight section of surplus WR-2300 waveguide (aluminium) with the inside dimensions of 584 x 292 x 1000 mm. Its ends have been closed by custom made endcaps (stainless steel). These are made from a frame with mounting holes, matching the waveguide flange. A perforated piece of sheet metal closes the waveguide while allowing an airflow for cooling the internal components. The sheet metal provides good mechanical stability – but its perforation holes of s = 3 mm in diameter significantly weaken the shielding performance, as predicted by:

$$SE_{\mathrm{perf}} \approx 20 \log \frac{\lambda}{2s}, \tag{3.10}$$

where λ is the wavelength of the EMI involved [81]. The equation predicts $SE_{\mathrm{perf}} \approx 24$ dB for the aforementioned metal sheet at 3 GHz. Shielding effectiveness was measured to be $SE_{\mathrm{perf}} \approx 30$ dB at this stage of construction.

To improve shielding performance, a thin wire mesh was applied on each side of the stainless steel plate with conductive glue. It consists of woven bronze wire with a mesh spacing of $s = 0.1$ mm. Equation 3.10 predicts, that a single layer of the mesh will improve shielding effectiveness to $SE_{\mathrm{perf}} \approx 54$ dB. Measurements showed $SE_{\mathrm{perf}} \approx 60$ dB at this stage of construction. After applying a second layer of wire mesh on the inside walls of the lids we could measure > 100 dB of attenuation. This provided sufficient shielding performance to avoid a potential bottleneck by the perforated waveguides lids. The various construction steps and the corresponding measured SE values are shown in Fig. 3.13.

To ensure a continuous conduction path between the waveguide and the two endcaps, gaskets made from knitted copper beryllium (CuBe) wire mesh have been applied with conductive glue to the waveguide flanges. Because of its high conductivity and low inductance, CuBe provides the best performance of any metal spring material [81]. Another advantage of CuBe is its hardness compared to oxidation films on the mating surfaces. With enough mating pressure it can form durable and self cleaning contacts, even after the shielding enclosure has been opened and closed many times.

A potential weak point of any electromagnetic shielding enclosure are cavity resonances, which can exist at certain frequencies, determined by the dimensions of the enclosure. They can lead to substantially reduced

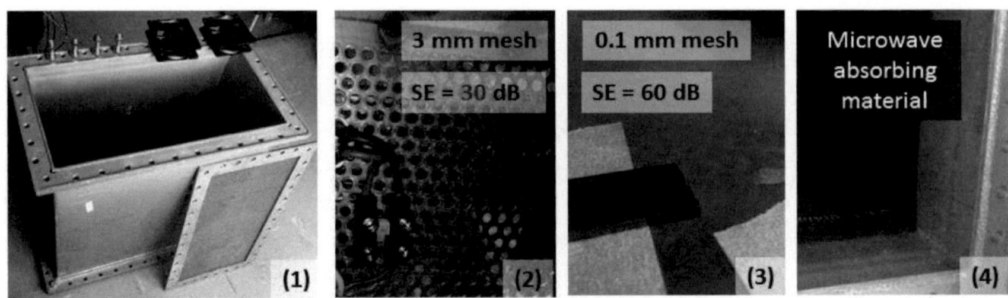

Figure 3.13.: Mechanical construction of the receiver shielding box. (1) WR-23 waveguide which serves as the base for the shielding enclosure. (2) Stainless steel plates with 3 mm holes as a endcaps on each side of the waveguide, allowing forced air cooling of the shielded instruments. (3) Fine metal mesh to improve SE of the lids. (4) Microwave absorbing foam to dampen cavity resonances.

shielding performance at these frequencies. The enclosure has been lined with microwave absorbing foam on its inside walls, which dampens cavity resonances, effectively mitigating the degradation of SE [82].

3.4.3. Construction of the cavity shielding box

The cavity shielding enclosure, shown in Fig. 3.12b, contains the detecting cavity and the front-end components, which are: low noise amplifier (LNA), bandpass filter and analog optical transmitter. For ALP measurements, the enclosure will be placed in the bore of a superconducting magnet. Therefore it needs to be as compact as possible and entirely constructed from non-ferromagnetic materials. This is to minimize distortion of the static magnetic field and to prevent strong attractive forces during insertion of the enclosure into the magnet[2]. The vessel has been machined from aluminium, the lid from brass. All screws are brass or nylon.

A technical drawing of the enclosure is shown in Fig. 3.14. Two grooves have been machined in the mating surface for the lid. This allowed us to

---

[2]The particular magnet which was used for the ALP run could not be ramped down.

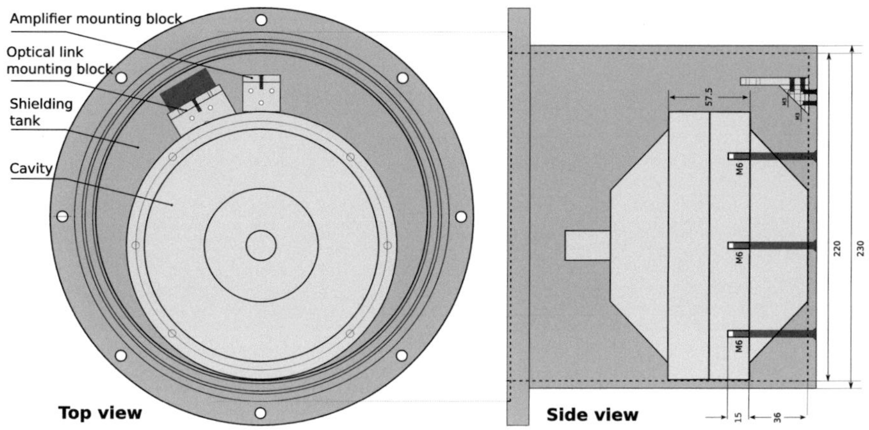

Figure 3.14.: Mechanical drawing of the shielding enclosure for the detecting cavity and front-end components. All dimensions are in [mm].

apply two EMI gasket made from CuBe wire mesh, of the same type as described in 3.4.2. Mounting blocks for the front-end components were machined from aluminium. The front-end dissipates up to 10 W of power and the mounting blocks act as heatsink, providing a path of low thermal resistance to the bulk material of the shielding enclosure. Components have been mounted to the bottom surface of the enclosure with countersunk machine screws. This requires through-holes in the walls of the shielding enclosure, which are – at first glance – potential weak spots for EM leakage. Experience showed that the countersunk screw heads in combination with the 5 mm thick walls provide adequate protection against EMI and no leakage could be measured in vicinity to the mounting screws.

For diagnostic purposes, a sinusoidal signal of known frequency is emitted within the shielding enclosure during each measurement run. This "test tone" of relatively low and constant power (≈ -100 dBm) couples from a $\lambda/4$ antenna to the detection cavity and to the components of the receiver front-end. By identifying the signal in the recorded spectrum, it was demonstrated that the entire signal processing chain was operational during the measurement. It is also possible to diagnose unwanted frequency offsets, frequency drifts or phase noise by comparing the shape and position

(a) EMI gaskets

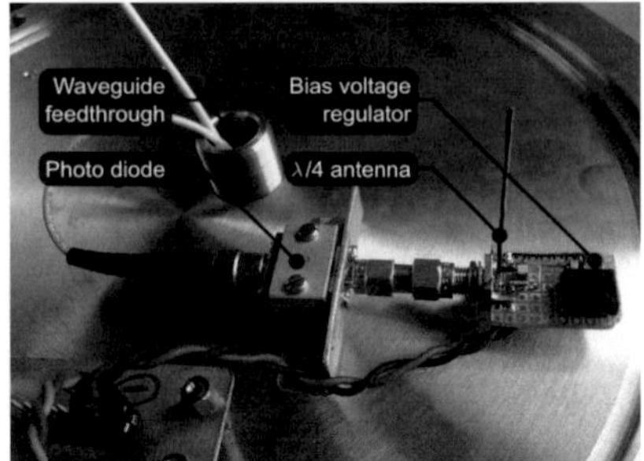

(b) Test tone transmitter

Figure 3.15.: The lid of the cavity shielding box can be removed. Two knitted wire gaskets made from CuBe provide uninterrupted flow for RF currents, blocking EMI. A test tone transmitter allows online diganostics of the shielding performance and of the RF receiver chain.

of the measured signal peak to the expected one. Furthermore, observing the test tone signal with constant power over time indicates that the EM shielding has not deteriorated. The test tone was transmitted over an optical fibre into the shielding enclosure, where a reverse biased photo diode (Hamamatsu G9801) converts the optical signal to an electrical one. The test tone frequency f_{test} has been offset by ≈ 400 Hz to f_{sys}, which avoids any interference with WISP detection.

3.4.4. Feedthrough filters

Each shielding enclosure is equipped with two different kinds of feedthrough filters. Their functions are:

Power feedthroughs: Power needs to be provided to the active components in both shielding enclosures without compromising the shielded domain. This is not straightforward, as EMI signals can readily propagate on copper cables. Furthermore, copper cables act as antennas, picking up EMI from the environment. The function of the feedthrough filter is to conduct DC or 50 Hz AC power into the shielding enclosures, while any EMI signals of higher frequencies are blocked. The stopband attenuation should be > 100 dB, to avoid a bottleneck for the shielding effectiveness of the entire enclosure system.

Waveguide feedthroughs: All analog and digital signals, in and out of each shielding enclosure, are transmitted over optical fibres. The waveguide feedthroughs are basically apertures in the shielding with well defined dimensions. They allow to feed fibres into the shielded domain, while preventing outside EMI to enter the enclosure.

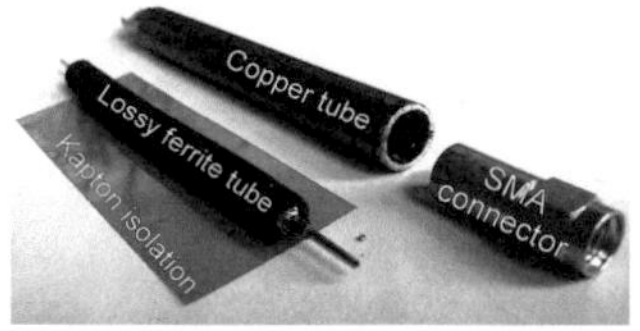

(a) Inductive element

(b) Assembled feedthrough filter

Figure 3.16.: Design of the 230 V feedthrough filter for powering the instruments in the receiver shielding enclosure. It is based on an inductive element containing a ferrite absorber and a commercial feedthrough capacitor, forming a second order lowpass filter.

The first filter to be constructed was the power feedthrough for the receiver shielding enclosure. It can be seen in Fig. 3.12a. The filter is required to feed 230 V AC power at 50 Hz into the shielded domain while providing > 100 dB shielding effectiveness against EMI at > 1 GHz.

The filter was custom made from three components: A commercial line filter blocks low frequency interference (< 10 MHz). Two commercial feedthrough capacitors and two custom made inductive elements block and absorb high frequency interference (> 10 MHz). The capacitors are of type SFCMC5000154MX (C=150 nF) manufactured by the company Syfer. High frequency signals are short circuited to ground by their very low reactance (10 $m\Omega$ at 100 MHz) while DC and low frequency AC signals are not affected (21 $k\Omega$ reactance at 50 Hz). As no specifications and no data on their performance at 3 GHz was available, they were measured in a test fixture as shown in Fig. 3.17. After the fixture was closed with a lid, shielding effectiveness of the filter could be derived from a S_{21} measurement

Figure 3.17.: Test fixture to measure the Shielding Effectiveness of the capacitive feedthrough filter from Syfer.

with a VNA. The measurement results for a similar filter (C=680 nF) are shown in Fig. 3.18. They indicate that a single capacitor provides $SE > 95$ dB.

While the 680 nF capacitor provides good EMI attenuation, it does not provide sufficient voltage ratings for reliable operation at 230 V, 50 Hz. Furthermore, the reactive current flowing to ground due to the filters capacitance ($I_r = 50$ mA) would be excessive. Therefore the 150 nF version had to be used ($I_r = 10$ mA), which can only achieve the design requirements through the addition of inductive elements. These are coaxial structures made from an inner conductor (copper), an intermediate tube consisting of lossy ferrite material and an outer conductor (copper tube). The structure is shown in Fig. 3.16a. It is sealed by SMA connectors at each end, which allows us to accurately measure the elements attenuation with a VNA (we measured SE $\approx$ 70 dB at 3 GHz, as shown in Fig. 3.18).

The requirements for the power feedthroughs of the cavity shielding box are less stringent as all of the electronics are powered from a single 15 V DC rail. The filters can be seen in Fig. 3.12b. The low DC voltage allows us to use the more effective 680 nF feedthrough capacitors (Syfer SFJNC2000684MX1). They have been utilized in a single stage, which already provides the required shielding performance. The inductive elements cannot be used for the cavity shielding enclosure, as it will be placed in a strong magnetic

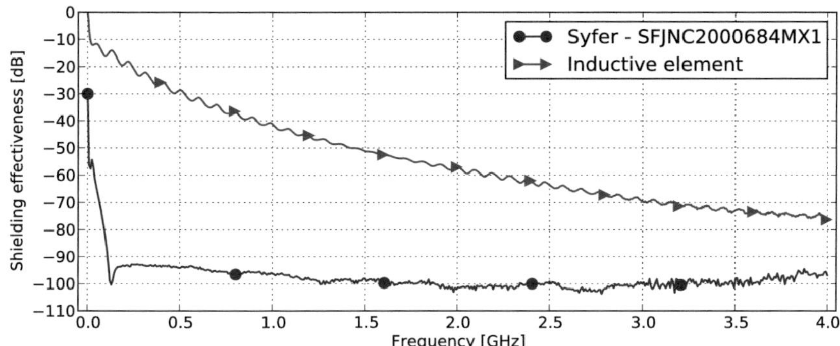

Figure 3.18.: Measurement of the Shielding Effectiveness (SE) with a Vector Network Analyzer. The inductive element was measured directly at its SMA connectors, the capacitive one in a custom test fixture (see Fig. 3.17). The final performance of the filter is the sum of both SE values.

field during the ALP runs. The ferrite material would saturate and lose its absorptive properties.

The second kind of filters, waveguide feedthroughs, have been used in both shielding enclosures to feed optical fibres in the shielded domain without compromising it. The feedthroughs can be seen in Fig. 3.12 and Fig. 3.15b. The feedthroughs are short brass or stainless steel pipes of diameter d and length l, penetrating the shielded domain. The pipes can be treated as circular waveguides: Electromagnetic interference below the cut-off frequency (f_c) cannot propagate and is attenuated exponentially [83]. The estimated screening effectiveness of the feedthroughs has been derived in [81]:

$$ f_c \approx \frac{176 \text{ GHz mm}}{d \,[\text{mm}]} \qquad SE \approx 20 \log \frac{f_c}{f} + \frac{27.3 \, l}{d} \qquad (3.11) $$

The utilized brass pipes are of $d = 10$ mm diameter and $l = 50$ mm length. The expected attenuation at $f = 3$ GHz is $SE \approx 152$ dB. While this provides a simplistic but effective feedthrough filter for optical fibres, copper cables cannot be fed through this opening. They would form a coaxial structure with the outer pipes. As TEM waves can propagate on coaxial structures, there would be no cut-off frequency and EMI could leak into the shielding.

3.4.5. Measurement of shielding effectiveness

Measurements have been carried out to determine the shielding effectiveness of the receiver shielding box, of the cavity shielding box, of the two cavities and of the VSA. The following devices have been used for the measurement: calibrated magnetic near-field probe (HP 11940A), low noise preamplifier (HP 87405A), VSA signal analyzer (Agilent N9010A) and RF signal source (Anapico 6010).

The near-field probe with the preamplifier, in combination with the VSA, operated on a narrow resolution bandwidth ($BW_{\mathrm{res}} \approx 0.1$ Hz) allows us to measure the strength of magnetic near fields with very high sensitivity. In this configuration, the probe antenna is also very useful to localize weak spots in the shielding enclosure under test.

Shielding effectiveness of the individual components has been measured and the results have been illustrated in Fig. 3.19. Due to the different nature of these components, different measurement variations had to be applied:

Receiver shielding box: The size of the enclosure allowed us to operate the RF signal source from inside of it. A wire of length $\lambda/4$ on the RF generators output port serves as antenna, radiating a test signal within the enclosure. The field strength of this signal at several fixed points outside of the enclosure has been measured with the near-field probe, preamplifier and VSA. This has been done for the open and closed enclosure. The shielding effectiveness can be determined from the ratio of these two measurements by Eq. 3.7 ([83]). Special care had to be taken taken, to prevent saturation of the preamplifier during the "open" measurement.

Cavity shielding box: The enclosure is too small for placing the RF source inside of it, therefore it has been measured in an inverted fashion. The probe antenna was connected to the RF source and used as a signal emitter, allowing us to produce a well defined field strength at certain fixed points outside the shielding enclosure (with about 10 cm distance to it). The front-end components (see 3.5) have been mounted and operated within the shielding box. Instead of the detecting cavity, a $\lambda/4$ antenna has been connected to the input port of the LNA. The signal power from the front-end has been monitored by the VSA. The ratio of the received power on the test tone frequency, with the shielding box opened and closed, allows us to determine the shielding effectiveness by Eq. 3.7.

Cavity: No components can be placed within the cavity's volume. Therefore the TE_{011} mode has been excited by the RF source on its resonant frequency. With the known RF input power P_{in}, the known loaded quality factor Q_L and with the assumption of critical coupling, the electric and magnetic field strength within the cavity can be estimated. The field strength outside the cavity has been measured with the combination of near field antenna, preamplifier and VSA. The ratio of magnetic field strength in- and outside of the cavity allows us to determine its shielding effectiveness.

To estimate the field strength within the cavity, the electric and magnetic field distributions of the cavity mode have been extracted from a CST microwave studio©simulation. The magnetic field (B_0) is normalized to 1 J of circulating energy and can be rescaled to an absolute field strength (B) by:

$$B = \frac{B_0}{\sqrt{Q_L P_{in}}} \tag{3.12}$$

VSA: The shielding effectiveness of the VSA has been estimated by terminating its input port with a matched load. A strong (30 mW) test signal from the RF source was emitted by the probe antenna and applied to different points on the enclosure of the instrument. A small amount of RF power could be detected by the instrument because of leakage through its casing.

No reference measurement was possible, therefore antenna factors and propagation constants have been neglected. Shielding effectiveness

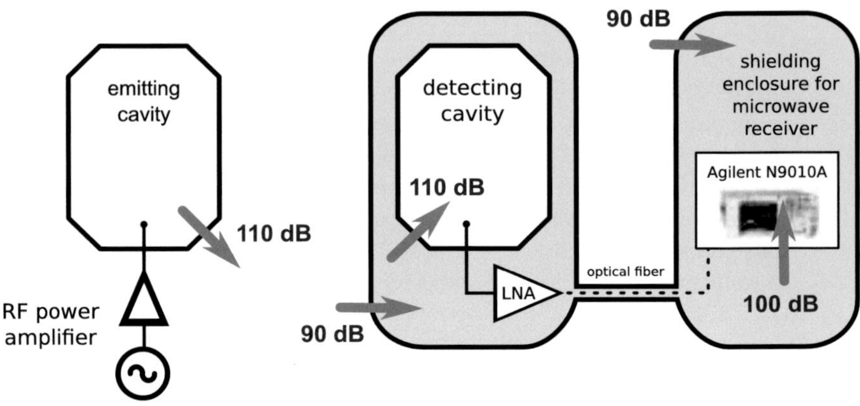

Figure 3.19.: Measured values for the shielding effectiveness of the two cavities, the two shielding enclosures and the VSA.

has been estimated from the RF power driving the probe antenna and the received RF power from the VSA.

Both enclosures provide $SE \approx 90$ dB and each of the cavities provides an additional $SE \approx 110$ dB of shielding. The combined shielding effectiveness of the experimental setup is therefore $SE_{sys} \approx 310$ dB, exceeding the requirement which was set in the beginning of 3.4 ($SE_{sys} \geq 260$ dB). Additionally the shielding provides a large enough performance margin to ensure that thermal noise is the limiting factor for the minimum detectable signal, even after opening and closing the enclosures many times.

3.5. Microwave front-end

The function of the microwave front-end is to amplify the noise like signal from the detecting cavity, apply a coarse frequency selection to reject out of band signals and to modulate the signal on an optical carrier, which is transmitted over an optical fibre to the shielding box outside the magnet.

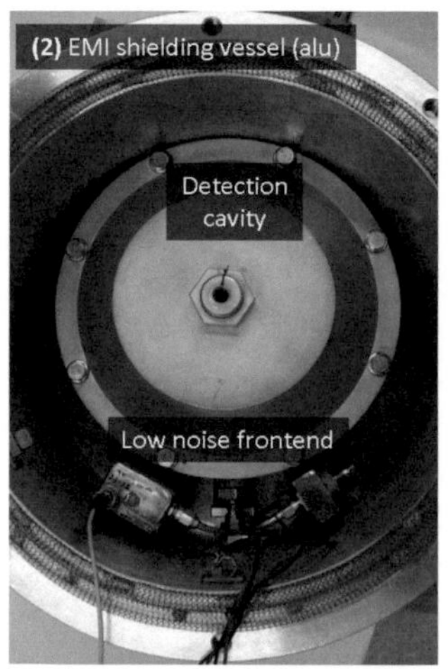

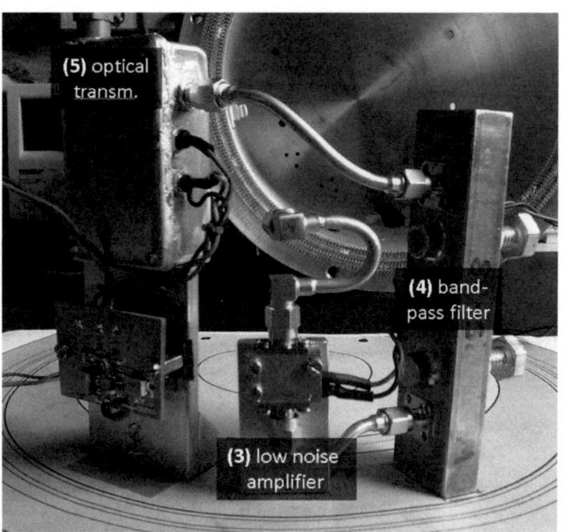

(a) Detecting cavity and front-end mounted in shielding enclosure.

(b) front-end outside the shielding enclosure. Components are connected by semi-rigid coaxial cables, which provide excellent screening attenuation.

Figure 3.20.: The experimental inset for the magnet contains the detecting cavity and a microwave front-end, consisting of: (3) low noise amplifier, (4) bandpass filter and (5) analog optical link.

The microwave front-end consists of three modules: a Low Noise Amplifier (LNA), a bandpass filter and an analog optical transmitter. A photo of the three components is shown in Fig. 3.20b. The LNA needs to be placed as close to the detecting cavity as possible to avoid degradation of the signal to noise ratio. Therefore all components of the RF front-end are mounted within the cavity shielding enclosure and need to be compatible with the 3 T magnetic field.

3.5.1. Low noise amplifier

The front-end and the components of the receiving chain further downstream of it are required to add as little additional noise as possible to the signal from the detecting cavity. In that sense, the first amplifier in the receiving chain is critical – if it provides sufficient gain, it will be the only component which will add significant additional noise to the received signal. Therefore, we demand that the noise temperature of the LNA is substantially lower than that of the detecting cavity, which is equal to $T_{nCav} \approx 300$ K.

A second requirement for the LNA is to provide enough gain so that the added noise from subsequent microwave components can be neglected. This is a consequence of the Friis formula (Eq. 7.20), indicating that if the gain of the first amplifier is much larger than the noise figure (NF) of subsequent components ($G_{LNA} >> NF_{comp}$), the amplifier will determine the noise performance of the entire receiving chain. This has been demonstrated in Fig. 3.26, where the noise figure of each individual component and the one of the complete receiving chain has been measured.

The selected LNA is a commercial amplifier from the company MITEQ (AMF-3F-0200-400-06-10P). It is based upon a feedback design approach, where each amplification stage is implemented by a single Field Effect Transistor (FET). Impedance matching is achieved by means of a feedback path between in- and output ports of the amplifier. A design trade-off between low noise figure and low input VSWR was carefully chosen by the manufacturer. The device operates at room temperature and is specified to provide > 40 dB of gain in a frequency range of 2 - 4 GHz, with a noise temperature of $T_{nLNA} \approx 35$ K (NF ≈ 0.5 dB). With this amplifier, the thermal noise from the detecting cavity is the only significant noise source in the receiving chain, as shown in Fig. 3.26.

Note that lower noise temperatures are possible with cryogenically cooled amplifiers. These devices are commonly used in radio astronomy and achieve noise temperatures in the order of $T_n \approx 2$ K. However, to reduce background noise, the cavity would also need to be cooled to similar temperatures. The background noise power could be reduced by a factor of $\approx 100 = 20$ dB, which would improve the sensitivity to WISPs by a factor of $100^{1/4} \approx 3.2$ (according to Eq. 2.15 and Eq. 2.16). However – cryogenic technology has not been utilized in the CROWS experiment yet, as it would dramatically increase the complexity of the setup, especially for ALP measurements in a magnet.

The MITEQ LNA has been tested and characterized in the 3 T magnetic field, which was necessary, as the components in the amplifier can be

Table 3.2.: Measured performance of the LNA in a 3 T magnetic field.

G = 45 dB	NF = 0.62 dB	T_{LNA} = 44.5 K

affected by a magnetic field of that strength [84]. The gain and NF of the amplifier was measured by the Y-factor method with a calibrated noise source, as described in B.2. The measurement results are shown in Table 3.2 and indicate a slightly higher noise figure than specified in its datasheet. At the time of writing, the amplifier was operated for > 100 h in the magnetic field without any degradation of performance.

3.5.2. Bandpass filter

The optical transmitter was found to saturate during measurements of the gain and noise figure of the frontend, which required to connect an active noise source instead of the cavity to the first amplifier. The problem could be traced back to the LNA, providing gain over a very wide bandwidth of $BW > 3$ GHz. At this point, no filters were used in the front-end. The broadband noise signal ($T_H = 9461$ K $= 159$ dBm/Hz) was amplified by the LNA ($G_1 = 45$ dB) and the optical transmitter ($G_2 = 16$ dB), which lead to a large integrated noise power of $P_{out} = $ -3.1 dBm at the output of the optical link:

$$P_{out} = k\ BW\ T_H\ G_1\ G_2 = 0.493 \text{ mW} = -3.1 \text{ dBm}. \tag{3.13}$$

Its 1 dB compression point is given by $P_{1dB} = -11.5$ dBm, therefore it was driven into strong saturation during the noise measurements.

As a solution, a non-magnetic and adjustable bandpass filter was designed, built and inserted between the LNA and optical transmitter. The filter reduces front-end bandwidth to $BW = 25$ MHz and integrated output power of the optical link to $P_{out} = -24$ dBm, preventing this saturation and effectively trading bandwidth for dynamic range.

The filter is based on an evanescent mode design as described in [85]. The RF signal is coupled in and out of a WR90 waveguide by two stub antennas with SMA connectors. The cut-off frequency of this waveguide

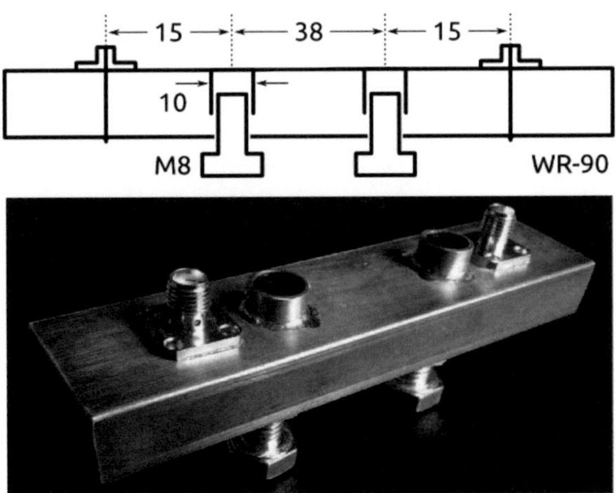

Figure 3.21.: Dimensions [mm] and photo of the magnetically compatible bandpass filter. The design is based on a section of WR-90 waveguide, which operates below its cut-off frequency [85]. The concentric tuning screws form capacitors, which are resonant with the waveguide and allow waves to propagate only at a specific frequency.

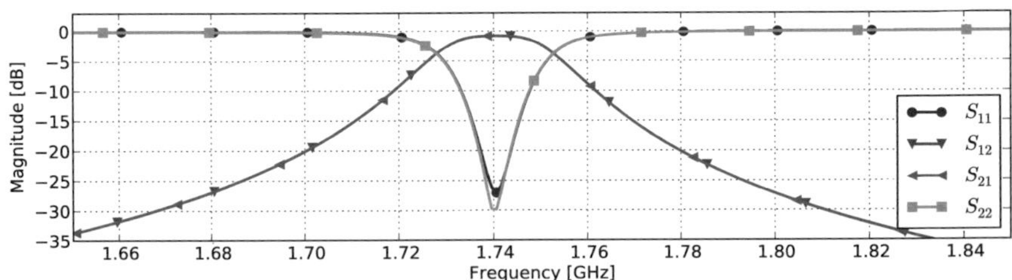

Figure 3.22.: Measured S-Parameters, showing transmission (S_{21}, S_{12}) and reflection (S_{11}, S_{22}) properties of the filter. A passband with a 3 dB bandwidth of $\approx$ 25 MHz and a insertion loss of 0.8 dB has been achieved. The center frequency is tunable in the range of 1.5 GHz - 3.5 GHz.

is $f_c = 6.56$ GHz, so direct signal propagation between the two stub antennas is not possible. Two adjustable capacitors, formed by brass screws in concentric copper tubes, are inserted into the waveguide. These capacitors resonate with the adjacent sections of waveguide, forming two weakly coupled resonators. They allow signals to pass between the two coupling stubs only at their resonant frequency. The critical parameters of the filter (the three distances between the capacitors and stub antennas) have been optimized by numerical simulations with CST microwave studio. After two design iterations (simulation and comparison to measurements), the desired performance of $BW_{3dB} \leq 25$ MHz and less than 0.8 dB insertion loss has been achieved.

3.5.3. Analog optical transmitter

Signals are transmitted over optical fibres between the two shielding boxes. This has two distinct advantages over coaxial cables in our application:

- They provide galvanic isolation and a nearly infinite screening attenuation. Microwave interference does not influence the optical carrier and cannot couple into the shielded domain.

- Optical fibres are free of metals, making them compatible with the waveguide cable feedthroughs [83] used in the two shielding enclosures (for details, see 3.4.4).

The HSP experimental runs up to 12/2012 have been completed with a commercial optical transmitter and receiver combination of type MITEQ LBT-50K4P5G-25-15-M14, shown in Fig. 3.23a. Subsequently, the front-end had to be upgraded to operate in strong magnetic fields, allowing us to search for ALPs. The MITEQ product was not specified for this kind of environment. However, no alternative device with sufficient performance and guaranteed magnetic compatibility could be found from industry. It was decided to test the optical transmitter module in the 3 T magnet, as a preparation for the ALP run in June 2013. The risk of eventual damage was taken into account.

During the test we observed a significant reduction in gain as the module was brought closer to the magnet. Within the 3 T region, the optical link was permanently damaged after a few seconds of operation. The

(a) The transmitter module on a mounting bracket.

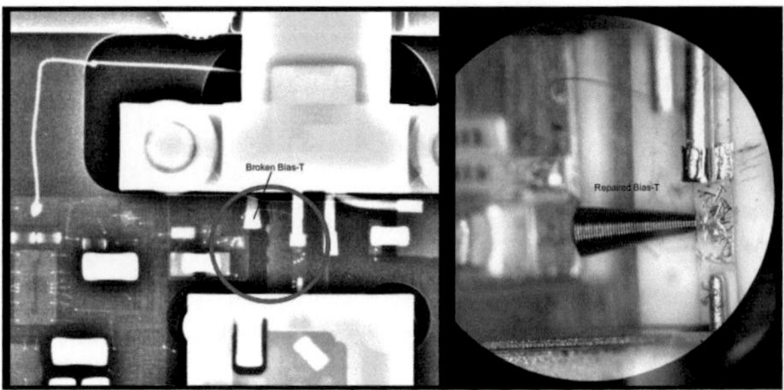

(b) X-Ray and microscope photos indicate the damaged component. A Bias-T was torn out from its fixation, as the bonding wires could not support the magnetic forces.

Figure 3.23.: MITEQ LBT-50K4P5G-25-15-M14 analog optical transmitter, which was permanently damaged in the 3 T magnetic field.

causes for these two modes of failure were found afterwards. The reversible reduction in link gain happened due to the presence of an optical isolator within the module. The component avoids reflected light from entering the laser diode and interfering with its operation. It does rely on the effect of Faraday rotation – a small permanent magnet within the isolator provides a well defined static magnetic field. By placing the module in the MRI magnet, a much stronger magnetic field was forced in the opposite direction, interfering with the isolators normal mode of operation. The permanent damage occurred after an internal component of the module was torn out from its mounting position. As Fig. 3.23b indicates, the damaged component was a Bias-T, containing a ferromagnetic core. The torque exerted on the component by the magnetic field was strong enough to snap its bonding wires.

As no magnetically compatible replacement product was readily available from industry, a commercial satellite TV Low Noise Block (LNB) from the company INVACOM was adapted for our purpose. This shortcut saved a significant amount of time, compared to designing an optical link from scratch. On the other hand, the circuit was cheap and uncomplicated enough to be customized for our purpose on board level. The 50 Ω mi-

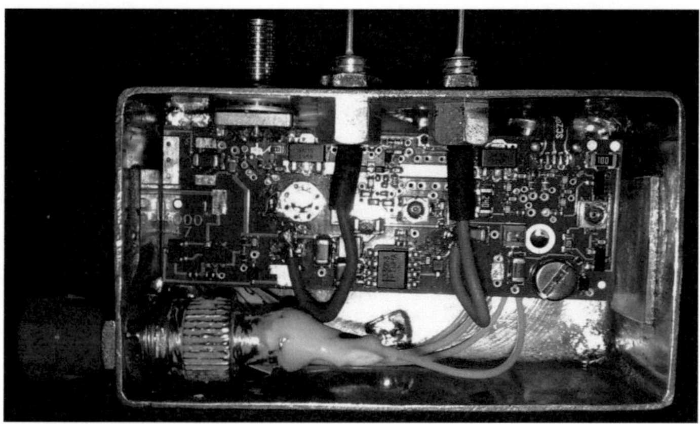

Figure 3.24.: INVACOM optical transmitter, modified for operation in
the magnet. Circuit board in an RF tight enclosure.

crostrip line, carrying the RF input signal to the transmitter was identified
and a SMA connector was directly soldered in place. The circuit board was
extracted and mounted in a custom, RF-shielded enclosure. Feedthrough
ports have been placed for two DC voltages, the optical fibre and the
SMA connector. Ferromagnetic materials of significant mass, including
all ferrite cored inductors have been removed from the circuit. These
components have been part of two internal DC/DC converters, which have
been replaced with a magnetically compatible power supply. For nominal
operation, $+6$ V and -5 V needs to be provided to the circuit. These
voltages were derived from the common 15 V rail of the shielding enclosure. A switched capacitor voltage inverter (LT1054) and linear regulators
(LM2941 / LM2991) were used for that purpose. The circuit board of the
custom power supply is visible in Fig. 3.20b.

The gain, noise figure and 1 dB compression point of the optical link were
measured, the results are shown in Table 3.3 and Fig. 3.25. Within the
frequency range of 0.5 - 3 GHz, comparable performance to the MITEQ
link was achieved. At the time of writing, the device operated reliably in
a magnetic field of up to 3 T for more than 100 h without showing any
signs of degradation.

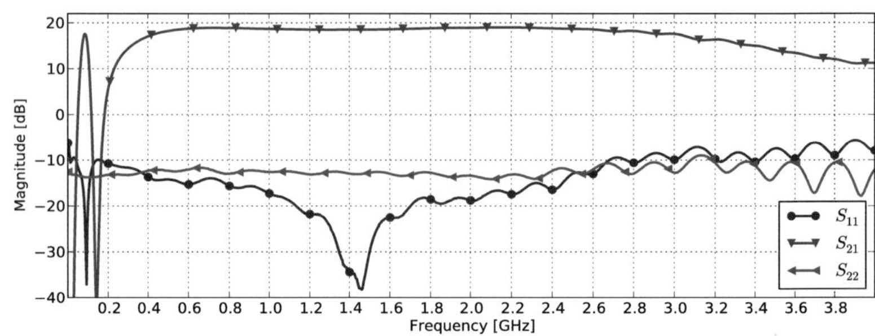

Figure 3.25.: Measured performance of the optical link, consisting of the INVACOM optical transmitter and the MITEQ LBR receiver. The link provides $\approx$ 17 dB of transmission gain at 1.7 GHz (S_{21}). The return loss of the optical transmitter input is $\approx$ -20 dB (S_{11}), indicating it is well matched to a 50 Ω system.

Table 3.3.: Comparison of the two optical transmitters at 1.74 GHz.

	Gain	Noise Figure	P1dB
MITEQ	16.0 dB	9.0 dB	-11.5 dBm
INVACOM	17.4 dB	10.6 dB	-10.0 dBm

3.5.4. Measurement of the Hardware Transfer Function

To be able to measure the power of a hypothetical WISP signal in case of a discovery, or to be able to determine the sensitivity of the experiment in case of an exclusion result, we are interested in the absolute noise power at the coupling port of the detecting cavity. However – this signal cannot be observed directly. What we actually record is the signal at the end of the receiving chain, modified by the attenuation, gain and added noise from all the components in the receiving chain. If the cascaded gain (G) and noise figure (NF) of all front-end components and cables between cavity and signal analyzer are known, we can calibrate the measured power spectra to obtain the power at the cavities coupling port. We will call the calibration values (G and NF) the Hardware Transfer Function (HTF).

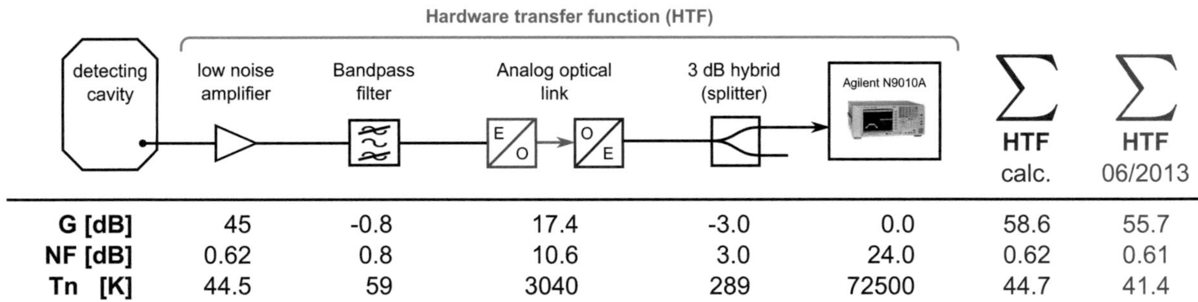

	detecting cavity	low noise amplifier	Bandpass filter	Analog optical link	3 dB hybrid (splitter)	Agilent N9010A	$\sum$ HTF calc.	$\sum$ HTF 06/2013
G [dB]	45	-0.8	17.4	-3.0	0.0		58.6	55.7
NF [dB]	0.62	0.8	10.6	3.0	24.0		0.62	0.61
Tn [K]	44.5	59	3040	289	72500		44.7	41.4

Figure 3.26.: Measured Gain (G), Noise Figure (NF) and noise temperature (T_n) of all microwave components along the receiving chain. **HTF calc.** is the Hardware Transfer Function, calculated from the individual component values by Eq. 7.20. **HTF 06/2013** is the result of the calibration measurement immediately before the ALPs run in June 2013. The discrepancy can be explained by the influence of the 3 T magnetic field on the components.

One option to determine the HTF is to measure each individual component and calculate the cascaded gain and NF with Eq. 7.20. This has been done and the result (measured outside the magnetic field) is shown in Fig. 3.26 in red. Note that this is an estimation of limited accuracy, as the attenuation of the coaxial cables and connectors is not taken into account and the final setup will be placed in the magnet which will influence the parameters of the components.

Another option is to measure the entire chain of components directly. This has been done for calibrating the experimental data, the results are shown in green. The HTF has been measured during the ALPs run in June 2013, immediately before recording data – after the experiment was placed in the magnet and warmed up. This measurement is more accurate, as it takes the attenuation from cables and connectors into account and minimizes errors due to thermal drift and due to the influence of the magnetic field.

The Y-factor method [86] was applied. For this purpose the experimental inset has been set up in the magnet. The detecting cavity was disconnected from the LNA and a calibrated noise diode (ENR = 15 dB) was connected through a 5 m long cable. The cable is necessary to prevent the noise diode from being influenced by the magnetic field. The cable's exact attenuation has been determined beforehand and taken into account. The principle behind the Y-factor method is described in B. The measurement has been carried out with the Agilent N9069A Noise Figure Measurement Application. For example, the HTF of the receiving chain during the ALP measurement run in June 2013 has been determined to be NF $= 0.6 \pm 0.2$ dB and G $= 55.7 \pm 0.5$ dB at f_{sys}. The measurement uncertainty has been estimated by a method described in [87].

As a concluding remark it shall be noted that one way to improve the usability of the present setup would be to automate the calibration measurement of the hardware transfer function. In the present state, the connection of the coaxial cable between the detecting cavity and the first amplifier needs to be opened manually. Then the HTF can be measured by connecting a calibrated noise source. This procedure could be simplified and sped up with a calibration tone of well defined amplitude, which is injected in the front-end by a directional coupler. Before the measurement, the calibration tone would be switched on, allowing us to determine the HTF in an automated way. One might even think of a periodic calibration during the measurement, allowing the receiver to operate as an radiometer, making it possible to accurately measure the noise temperature of the detecting cavity. An automatized calibration signal with the required precision and stability was not implemented in the current setup due to time constraints. Although in the current setup a test tone signal has been implemented, which could be used as calibration tone, its amplitude is not accurately enough defined. Moreover, instead of being coupled into the frontend with a directional coupler in a well defined way, it is radiated from an antenna, which makes the received signal dependent on its position and also the shielding of the components. The latter has been exploited, as one of the uses of the test tone is to diagnose potential problems with the shielding of the receiving chain.

4. Signal detection

4.1. RF receivers

The detecting cavity provides a microwave signal, which essentially consists of thermal noise, shaped by its resonant modes. The signal is amplified and filtered by the components of the front-end and subsequently transmitted over an optical fibre to the receiver shielding enclosure, which is placed several meters from the magnet. An analog optical receiver (MITEQ LBR-50K4P5G-10-15-10) converts the optical signal back to the electrical one. The spectral content of the signal at this point has been illustrated in Fig. 4.1. The significant part of the spectrum for WISP detection corresponds to a rather narrow slice around the frequency f_{sys}. The function of the microwave receiver is to select this slice, shift it down in frequency, convert it to the digital domain and record it for later processing.

An overview of the receiving chain, as it has been used during the ALP measurement run in June 2013 is shown in Fig. 4.2. Two receivers have been used in parallel: The primary receiver is a commercial Vector Signal Analyzer (VSA) of type N9010A from Agilent. The secondary receiver is a dedicated downmixing chain, optimized for low noise and frequency stable recordings. This section will focus on the primary receiver. The secondary receiver has been described in 4.6.

The basic signal processing steps within the VSA are shown in 4.3. The architecture of the instrument is very similar to a modern digital communication receiver. An analog tuner converts the incoming signal to an Intermediate Frequency (IF) in several steps, using mixers, local oscillators and filters to achieve this. The resulting IF signal is digitized by an analog to digital converter (ADC) with 16 bit resolution. The VSA applies a digital decimation filter to reduce the span of the frequency slice, before

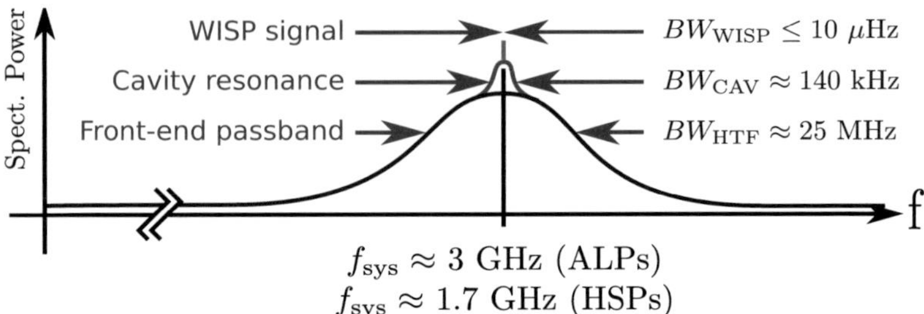

Figure 4.1.: Spectral power of the expected noise like signal from the front-end. The WISP signal would appear as a sinusoid of very narrow bandwidth at the frequency f_{sys}.

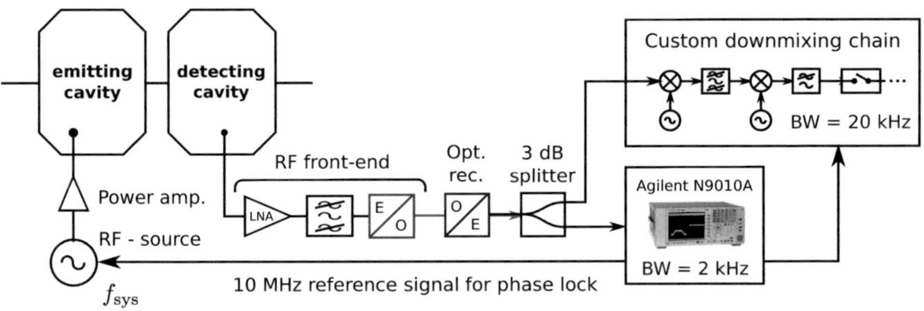

Figure 4.2.: Schematic of the RF components in the receiving chain. Two receivers record the WISP signal in parallel, which allows us to cross-check the results. This setup has been used for the ALP run in June 2013. For the HSP run in Sept. 2013, the custom downmixing chain and the 3 dB splitter was not used.

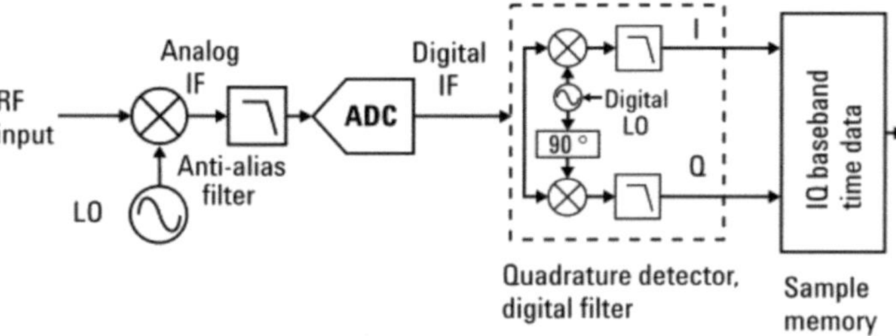

Figure 4.3.: Overview of the signal processing carried out by the Agilent Vector Signal Analyzer (from [88])

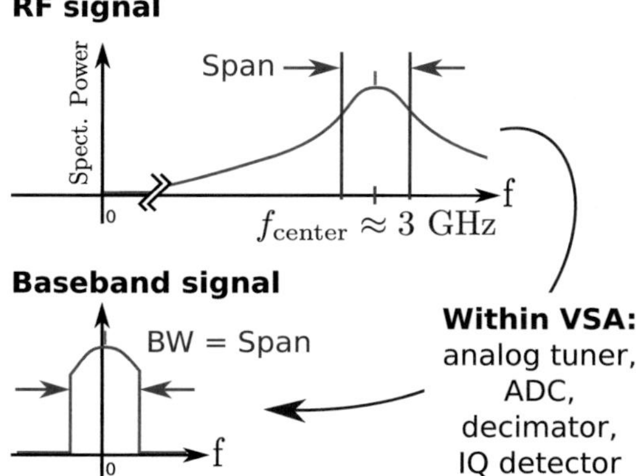

Figure 4.4.: Illustration of the signal processing in the VSA. A slice with a width corresponding to the span setting on the instrument is isolated and converted to a baseband quadrature signal of the same bandwidth.

it is further processed. The decimated IF signal is converted to a base-band signal by a quadrature detector. Each sample of the time domain base-band signal consists of a complex number with a real and imaginary part. It describes the modulation in amplitude and phase of a sinusoidal carrier of frequency f_{center} and thus can cover negative frequencies. The sample-rate of the base-band signal is higher by a factor of 1.28 than the chosen recording span, allowing the instrument to accommodate the transition band of its internal filters without aliasing. The base-band signal is stored in the acquisition memory of the analyzer. At the end of a recording, the memory is read out and the data is further processed by the integrated control PC. The Agilent application software carries out calibration and normalization of the raw IQ samples. Finally, the time domain data is stored as a binary file on a hard-disk. This stored time recording is the input data for all further offline analysis steps, which are described in 4.2.

The 2 GB acquisition memory of the VSA sets a limit on the maximum number of samples which can be acquired in a continuous fashion (N_{max}). Hence the product of recording span (SP) and recording length (l) is constrained to the following:

$$1.28 \cdot SP \cdot l \leq N_{max} = 268 \cdot 10^6. \tag{4.1}$$

For example, for a given recording time of 24 h, the span must be smaller than 2423 Hz.

This limitation is one of the reasons why a custom downmixing chain has been developed. It was designed to allow continuous data acquisition over a practically infinite time with a frequency span of up to 20 kHz. Moreover, it has been optimized for low noise, low phase noise and low frequency drift performance. This provides advantages over the VSA, which is a versatile but complex general purpose instrument, optimized for high dynamic range. Details about the custom double heterodyne receiving chain can be found in 4.6. For the ALPs run in June 2013, the signal from the detecting cavity was split by a 3 dB hybrid, which allowed us to operate both signal receivers in parallel to compare their results.

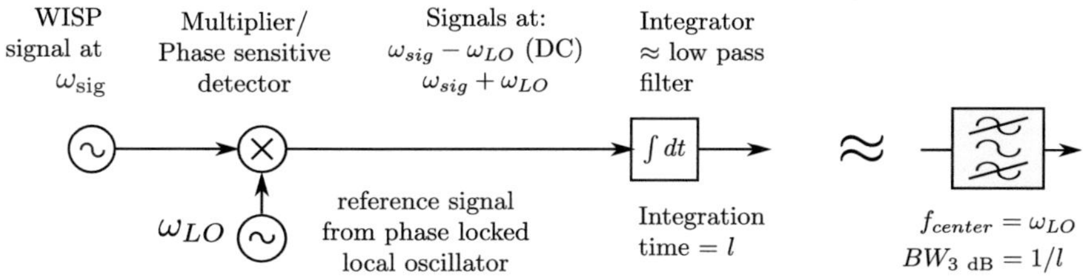

Figure 4.5.: An efficient way to detect a sinusoidal within white noise is the synchronous detector. The input is multiplied with a reference signal of the same frequency. Only if a signal exists within the noise, a non vanishing DC component can be observed at the output of the integrator. The synchronous detector can be compared to a bandpass filter.

4.2. Narrowband detection principle

The VSA stores the recorded time domain trace as baseband IQ data in a binary file. The complex signal has a bandwidth of $SP \approx 2$ kHz and consists to a good approximation of Gaussian white noise. The function of the signal processing is to determine the existence of an underlying sinusoidal signal – hidden within the noise.

One of the most efficient ways to detect "hidden" signals is a synchronous detector as illustrated in Fig. 4.5. It exploits the fact that the frequency of the WISP signal is known accurately and is equal to the excitation frequency of the emitting cavity (f_{sys}). The synchronous detector works by mixing (or multiplying) the input signal with a locally generated reference signal of identical frequency. Assuming that there is a weak WISP signal, the output of the mixer will show the sum and difference frequency between the WISP and the reference signal, which can be seen from trigonometric addition theorems. As they are both identical in frequency, the difference signal does not vary in time (DC). The product is fed to an integrator, which acts as steep lowpass filter. For long integration times, only the DC component can pass through the integrator, while periodic signals are suppressed. The whole detector device can be compared to a bandpass filter with a center frequency of f_{sys} and a passband which is proportional to the integration time $1/l$.

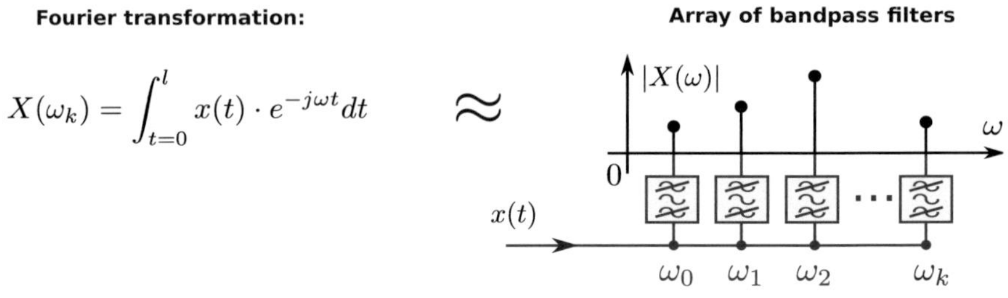

Fourier transformation: **Array of bandpass filters**

$$X(\omega_k) = \int_{t=0}^{l} x(t) \cdot e^{-j\omega t} dt \qquad \approx$$

Figure 4.6.: The Fourier transformation of a time domain signal can be interpreted as filtering the signal by an infinite number of synchronous detectors, tuned to successive frequencies ω. The bandwidth of each filter is inverse proportional to the length of the time trace l.

Keeping the principle of the synchronous detector in mind, one can inspect the definition of the continuous Fourier Transform (FT) in Fig. 4.6. It can be seen that the result of the FT $X(\omega_k)$ corresponds to a synchronous detection at the frequency ω_k: the input signal is multiplied by a complex phasor of that frequency and the resulting product is integrated over the measurement time l. If the operation is carried out for a large number of frequencies k, the spectrum of the input signal can be calculated.

In our case, the signal processing is done in the digital domain and carried out over a finite number N of discrete time domain samples, hence the Discrete Fourier Transform (DFT) is used. It is defined as:

$$X_k = \sum_{n=0}^{N-1} x_n \cdot e^{-j2\pi nk/N} \qquad k = 0, \ldots, N-1, \qquad (4.2)$$

where x_n is the n-th time domain sample and X_k is the k-th frequency bin. Note that in this case k is a normalized frequency, which relates to a frequency in [Hz] by Eq. 4.4. The continuous integration of the FT corresponds to a sum over a finite number of samples N in the DFT. The complex phasor is defined by its normalized frequency k. It can be seen that the DFT implements an array of digital synchronous detectors. Note that for sinusoidal input signals with frequencies corresponding to the spectral bins, the DFT is effectively implementing an array of matched filters [89] and therefore provides optimum detection sensitivity.

From a computational point of view, the most significant advantage of using the DFT is the availability of very efficient implementations based on the Fast Fourier Transform (FFT) algorithm. While the DFT requires $O(N^2)$ operations, the FFT can achieve the same results in $O(N \cdot \log_2 N)$ operations[1].

We make use of the DFT to determine the power spectrum of the recorded signal. For that purpose, a collection of scripts for the programming language Python have been developed. The estimated power spectrum by the Periodogram method is calculated by:

$$S(k) = \frac{|X(k)|^2}{N^2},\tag{4.3}$$

where k is the normalized frequency bin index. The corresponding vector of frequencies in [Hz] is given by:

$$f = \left[0, 1, \ldots, \frac{N}{2} - 1, \ -\frac{N}{2}, \ldots, -1\right] \cdot \frac{f_s}{N} \quad \text{if } N \text{ is even} \tag{4.4}$$

$$f = \left[0, 1, \ldots, \frac{N-1}{2}, \ -\frac{N-1}{2}, \ldots, -1\right] \cdot \frac{f_s}{N} \quad \text{if } N \text{ is odd.}$$

Where f_s is the sample rate in [Hz] and N is the number of recorded samples. Note how negative frequencies are arranged on higher indices than the positive frequencies. For displaying purposes, these two blocks are swapped, resulting in a frequency vector ranging from $f = -f_s/2 \ldots f_s/2$ for even spectra.

We define BW_{res} as the resolution bandwidth of the DFT:

$$BW_{\text{res}} = \frac{1}{l} = \frac{f_s}{N}.\tag{4.5}$$

---

[1] The FFT can only operate on input data of length $N = 2^a$, where a is a positive integer – but this limitation has been lifted in modern implementations like the highly optimized "Fastest Fourier Transform in the West" (FFTW) library. It achieves faster than $O(N \cdot \log_2 N)$ performance in most cases, even for arbitrary sized input data.

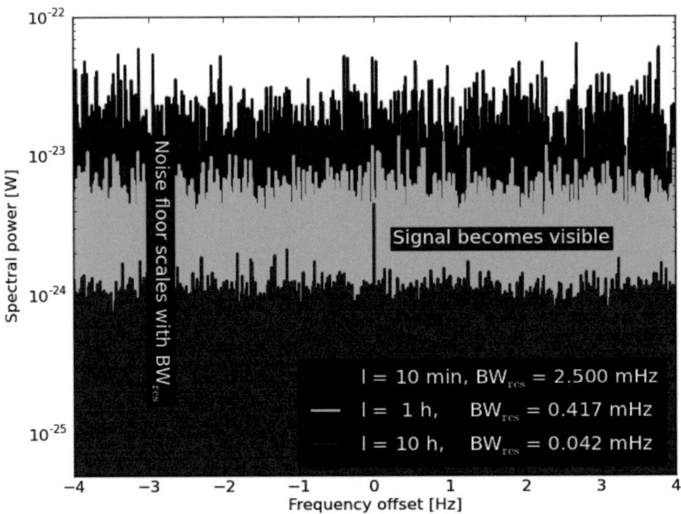

Figure 4.7.: Demonstration of the narrowband detection concept. The noise floor decreases with longer measurement times (narrower BW_{res}). A potential signal stays at constant power level. The plot has been derived from actual measurement data. The signal peak has been introduced by deliberate RF leakage.

Each bin of the power spectrum $S(k)$ corresponds to the integrated output power of a bandpass filter, tuned to a center frequency given by Eq. 4.4 and with a bandwidth equal to BW_{res}.

The displayed average background noise level in each spectral bin is given by

$$P_n = BW_{\mathrm{res}}N_0 = kBW_{\mathrm{res}}T_n, \qquad (4.6)$$

where N_0 is the noise power density in [W/Hz] of the input signal. Equivalently, the background noise can be described by its noise temperature T_n. A pure sinusoidal signal at fixed frequency (like the one we would expect from a WISP) has an infinitely narrow bandwidth and will always deposit its entire signal power (P_{sig}) within one spectral bin and hence within the passband of one virtual bandpass filter. Therefore the signal to noise ratio

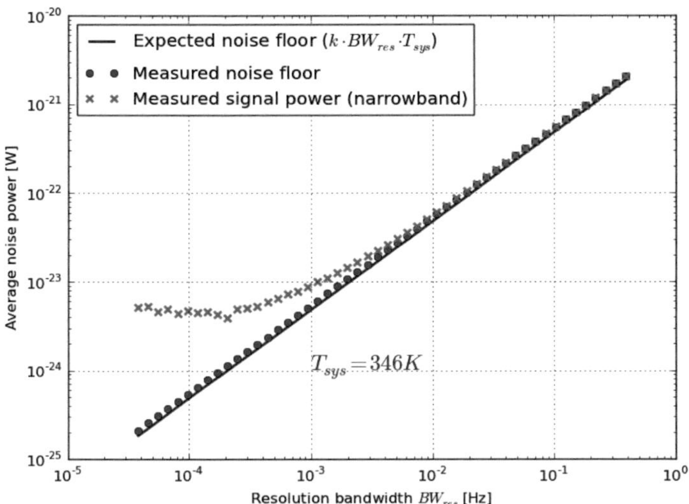

Figure 4.8.: RMS noise floor and signal power as a function of resolution bandwidth, demonstrating how the relevant frequency bin becomes dominated by the signal power once the resolution bandwidth becomes narrow enough.

in this bin, defined as $\mathrm{S/N} = P_{\mathrm{sig}}/P_n$, is proportional to the length of the recorded time trace. This is in contrast to averaging n spectra, where $\mathrm{S/N}$ only improves by a factor of $\sqrt{n}$.

This noise reduction effect has been demonstrated by evaluating recorded time traces of increasing length l. The resulting power spectra are shown in Fig. 4.7, exhibiting lower noise levels for longer recordings with narrower BW_{res}. For the longest recording length, a sinusoidal signal becomes visible as a peak emerging from the noise floor. This effect can also be observed in Fig. 4.8, where average noise power and power of the signal peak have been plotted as a function of BW_{res}.

So far we have shown the spectral data on the scale of a power spectrum with the unit [W]. This allows us to correctly read the power of a sinusoidal signal peak, independently of the resolution bandwidth of the spectrum. On the other hand, this scale cannot be used for noise like signals, which are described by a power density with the units [W/Hz]. Nonetheless, by normalizing the power spectrum $S(f)$ in the following way:

$$\mathrm{PSD}(f) = S(f)/BW_{\mathrm{res}}, \tag{4.7}$$

the power spectral density PSD can be calculated. This normalization is useful to characterize noise like signals, as the average value of the noise floor becomes independent of BW_{res}.

4.3. Properties of the Periodogram

So far it has been shown that synchronous detection is a good way to discover sinusoidal signals in white noise. It has also been shown that the power spectrum of a signal can be estimated by the Periodogram method, where each of the resulting spectral bins corresponds to a synchronous detector tuned to a specific frequency, which allows to evaluate a wide frequency range efficiently for signal peaks. Nonetheless, the Periodogram is only an estimate of the underlying power spectrum and suffers from distortions and certain artifacts, originating from inherent properties of the DFT. These unwanted effect and their influence on signal detection shall be discussed here. The most important ones are:

Spectral leakage: Sharp peaks in the spectrum are smeared out and appear to be framed by sidelobes, limiting the dynamic range and making it harder to distinguish between two signals if they are close in frequency to each other. More details on this effect are given in 4.3.1.

Scalloping loss: If a sinusoidal signal falls exactly between two frequency bins, its amplitude can be reduced by up to a factor of $2.5 = 3.9$ dB [90]. Further details on this effect are given in 4.3.2.

Variance: For noise like input signals, each frequency bin fluctuates statistically. Its variance is in the same order as its mean value, resulting in a very broad noise floor. In 4.3.3 we investigate whether reducing this variance by averaging is advantageous for the detection of small signal peaks.

4.3.1. Spectral leakage

In 4.2, a DFT frequency bin has been compared to a bandpass filter. Now the response function of that frequency bin shall be examined in detail. If the time trace is sinusoidal and infinitely long, the response in the spectrum would be an infinitely narrow peak. However, the time series needs to be truncated, as the DFT operates on a limited number (N) of input samples. In the time domain, this can be seen as an implicit multiplication with a rectangular "windowing" function. In the spectrum, this corresponds to an implicit convolution of the ideal signal peak with a $\sin(f)/f$ function. This "response" function is shown as the blue trace in Fig. 4.9 on a linear and in Fig. 4.10 on a logarithmic scale. The figure illustrates what would be seen instead of a sharp peak in the time-truncated spectrum. Another complication is the fact that the DFT will not return a continuous function as illustrated in the figures but only a small number of sampling points from this underlying response curve. The sampling points are illustrated in Fig. 4.9 on a linear scale. Note that the location of the response curve with respect to the DFT sampling points is determined by the frequency of the sinusoidal signal. If the frequency is an integer multiple of the resolution bandwidth (BW_{res}), the response curve is exactly centered on one of the DFT sampling points (as shown by the blue trace in Fig. 4.9 and Fig. 4.10). In this special case, no spectral leakage would be visible in the DFT result because only the main lobe of the response curve is sampled at a non-zero value – all other sampling points fall exactly on the zeros.

In the CROWS experiment, the frequency of the RF sources cannot be controlled with sub-BW_{res} resolution, therefore the more general case needs to be considered, where the sinusoidal signal has a fractional frequency relationship to BW_{res} and will not fall perfectly into a DFT bin. In this case, the response curve can be shifted by up to ± 0.5 bins with respect to the nearest DFT sampling point. The worst case scenario is shown by the green trace in Fig. 4.9. Now the side-lobes of the response function are no longer suppressed and become visible in the frequency bins calculated by the DFT. Furthermore, the full amplitude of the main lobe is not captured anymore – this apparent attenuation is called scalloping-loss and treated in 4.3.2. The side-lobes cause the signal power to leak into adjacent frequency bins, which makes it harder to distinguish between two neighboring signals.

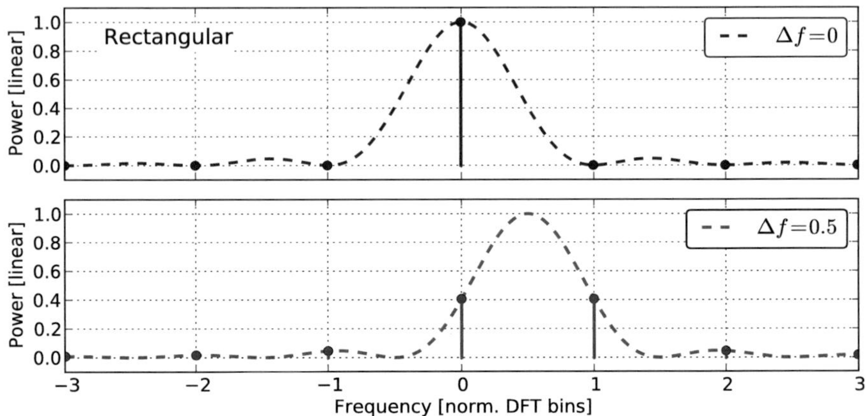

Figure 4.9.: Frequency response of the DFT without a window function to a sinusoidal signal. The sampling points as returned by the DFT have been illustrated. Spectral leakage is suppressed and the correct peak amplitude is returned if the signal frequency is an integer multiple of $\mathrm{BW_{res}}$ (blue trace, $\Delta f = 0$). In the other case (green trace, $\Delta f = 0.5$) spectral leakage and scalloping loss are visible.

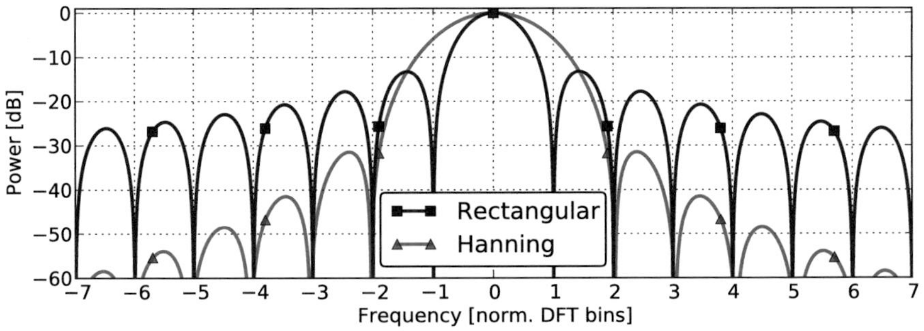

Figure 4.10.: Frequency response of the DFT without and with a Hanning window. Note how the side-lobes of the windowed DFT fall-off steeper at the expense of a wider main-lobe.

Small signals in vicinity of big ones are overshadowed and dynamic range is therefore limited.

Spectral leakage can be reduced by multiplying the time domain trace with a window function, which has been illustrated in Fig. 4.10 by the example of a Hanning window. Note how the response function falls off steeper and with less side-lobes than in the rectangular case – however, this is at the expense of a wider main lobe, a wider BW_{res} and therefore a higher average noise floor P_N.

For the CROWS experiment, the expected signal of a reconverted WISP is a single sinusoid with constant frequency. It is not necessary to resolve signals, which are tightly spaced in frequency or which have a large dynamic range. For making the decision, if there is a signal peak in the spectrum or not, spectral leakage is irrelevant. Therefore no window function was used before calculating the DFT. This yields the most narrow resolution bandwidth BW_{res}, the lowest average noise level in each bin (P_N) and the largest possible S/N for detecting sinusoidal signals [90].

Note that spectral leakage is however relevant for the test tone signal: In the final spectrum, its width is evaluated to identify any potential problems with the frequency stability of the RF sources. Therefore the minimum width of a spectral peak, as governed by the rectangular window, needs to be taken into account. As a visual aid, such an ideal peak has been overlaid with the test tone in Fig. 5.11.

4.3.2. Scalloping loss

If a sinusoidal signal falls between two frequency bins in the spectrum, its amplitude can be reduced by up to a factor of $2.5 = 3.9$ dB [90]. In the experimental run, it is not possible to control with sub-frequency-bin accuracy where a signal peak will fall. Moreover, in the very likely case of an exclusion result, there would be no signal peak visible at all. Nonetheless we need to provide an estimate about the sensitivity of the signal detector to determine the exclusion limit. Therefore it would be necessary to assume the worst case and take a reduction of the detection sensitivity by 3.92 dB into account. To avoid this trade-off, a solution needs to be found to reduce scalloping losses.

The amplitude inaccuracy happens due to the discrete nature of the DFT. As defined by Eq. 4.2, the DFT returns only N frequency bins from an underlying continuous spectrum: The absolute minimum amount of points from which the original signal can be reconstructed (corresponding to the Nyquist–Shannon sampling theorem in the frequency domain). These sampling points have been illustrated in Fig. 4.9.

Depending on the frequency of the signal, the peak of the continuous response function can be shifted by up to ±0.5 frequency bins from the next sampling point. Consequently, there is no guarantee that the maximum of the peak is captured by the DFT result. The apparent power of the signal is reduced by so called scalloping losses. This has been demonstrated in Fig. 4.11: The red curve is the result of a sinusoidal signal falling exactly between two frequency bins, leading to a degraded response in both bins.

One way to mitigate scalloping losses is to apply window functions to the time domain trace, as explained in 4.3.1. Especially flat-top windows have been optimized for high amplitude accuracy. Nonetheless, every window function will trade a wider BW_{res} and therefore a proportionally higher average noise floor for a more accurate amplitude response. A good trade-off regarding the worst-case sensitivity is the Welch window, reducing worst case scalloping losses to 2.2 dB while increasing the resolution bandwidth by a factor of 1.2.

A better way to reduce scalloping losses – which does not involve any trade-offs – is to define more than the critical number of sampling points and hence calculate more frequency bins. The tighter spacing of sampling points will capture the true maximum of the continuous response function much more accurately and hence scalloping losses are effectively reduced. This kind of oversampled DFT can be realized i.e., by zero padding the time domain trace. For large zero padding factors, the spectrum will approach the underlying continuous frequency domain function. This is demonstrated in Fig. 4.11. The green and red traces show the response of 7 spectral bins on a linear scale as calculated by the DFT. A signal is falling between two frequency bins and scalloping losses can be observed: the response does not reach the full amplitude. For the grey curve, the time domain signal has been zero-padded by 10x of its original length before the DFT operation. Scalloping losses are mitigated and the correct

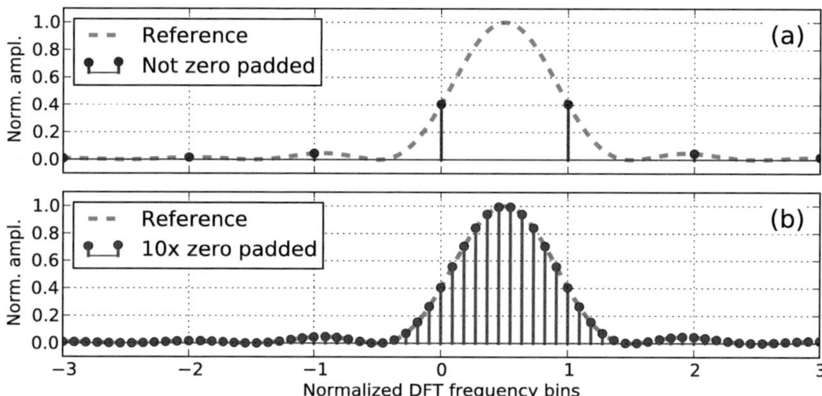

Figure 4.11.: Demonstration of scalloping loss. If a signal peak falls between two DFT bins as in (a) with $\Delta f = 0.5$, its apparent amplitude is reduced by up to 3.92 dB, if no zero-padding is applied. The loss in amplitude can be restored by oversampling the DFT spectrum, which can be achieved by zero-padding the time domain data. Figure (b) demonstrates how a 10x zero-padded DFT allows to significantly reduce the amplitude error.

amplitude is returned. Note, that the result is an interpolated spectrum with the same noise floor – zero-padding does not influence BW_{res}.

This method has been applied for the experimental data evaluation. The time traces have been zero padded with 10 times the number of recorded samples. In the plots of the power spectra (Fig. 5.6), the zero padded result is shown as grey curve, which does not suffer from scalloping loss.

4.3.3. Variance

Our goal is to detect the presence of a weak sinusoidal WISP signal within a background due to thermal noise from the detecting cavity. In the power spectrum, the background noise will appear as statistical fluctuation of each frequency bin around a certain mean value with a certain variance. Within the recording span of $SP = 2$ kHz, we can assume that the noise is "white", hence its statistical parameters are the same for each frequency bin. The WISP signal would appear as an excess of power in a single bin of known frequency. The probability of detecting its presence is related to the statistical properties (mean and variance) of the background noise in that bin. While applying the DFT operation to a longer time trace will lower the mean noise power, it will not reduce its variance. This is a mathematical property of the DFT. On the other hand, the time trace can be split into several blocks which are separately processed by a DFT and then averaged. This will reduce the variance, resulting in a more well defined noise floor – but at the expense of a higher mean value. In this section we will discuss, which of the two methods provides advantages for signal detection.

The yellow trace in Fig. 4.12a shows the background noise of a typical measurement run. The corresponding histogram is shown in Fig. 4.12b. Both plots are on a logarithmic (dB) scale. The mean (or median) value of the noise power is defined by Eq. 4.6, marking the center of mass of the distribution, as indicated by a red line. The green trace in Fig. 4.12 demonstrates how averaging several blocks will make the distribution approach a gaussian, centered around P_n.

We define the 99.9 % percentile of the noise floor as a figure-of-merit to gauge the sensitivity of the experiment. Only 0.1 % of all measured

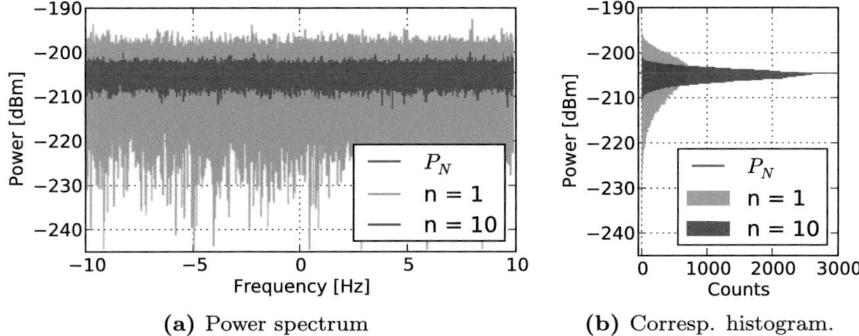

(a) Power spectrum **(b)** Corresp. histogram.

Figure 4.12.: Effect of averaging on the power spectrum of a noise like signal. n blocks of fixed length N have been averaged. The average noise power P_n has been marked as red line. Averaging reduces the variance in the spectral bins and makes them approach P_n. This is also evident from the histogram.

samples show a larger random fluctuation. Hence a WISP signal could be detected rather reliably, if it falls above the 99.9 % percentile.

To investigate whether averaging is advantageous for signal detection, the median and percentile values have been derived for power spectra, averaged over a different number of blocks. The result is shown in Fig. 4.13. Note how the decrease of the percentile while the mean noise power is not affected by averaging.

However, so far it has been assumed that a time trace of unlimited length is available. For example this can be the case in a measurement instrument, recording and processing data in realtime. This is in contrast to the goal of this work, where we want to find the optimum method to process a given time trace of limited length. Under these conditions, Welch's method is a popular way to perform spectral estimation. The time trace is split into several overlapping blocks, the power spectrum of each one is calculated and the results are averaged. A trade-off needs to be found: splitting the time trace in a larger number of blocks leads to a more defined spectrum – but at the expense of less samples (N) within one block, a wider resolution bandwidth (BW_res) and a higher average noise floor due to Eq. 4.6.

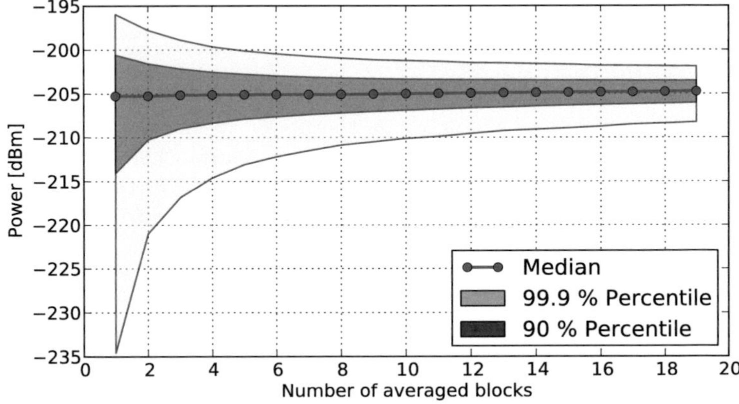

Figure 4.13.: For detecting sinusoidal signal peaks, not only the mean power, but also the variance of the spectral background noise floor needs to be considered. It has been visualized by its 99.9 % Percentile (yellow area). An increasing number of DFT results over a fixed blocksize have been averaged in this example, assuming that a time trace of arbitrary length is available. Note how the variance reduces with more averaged blocks, while the average noise power stays constant. This is a direct result of the fixed DFT blocksize.

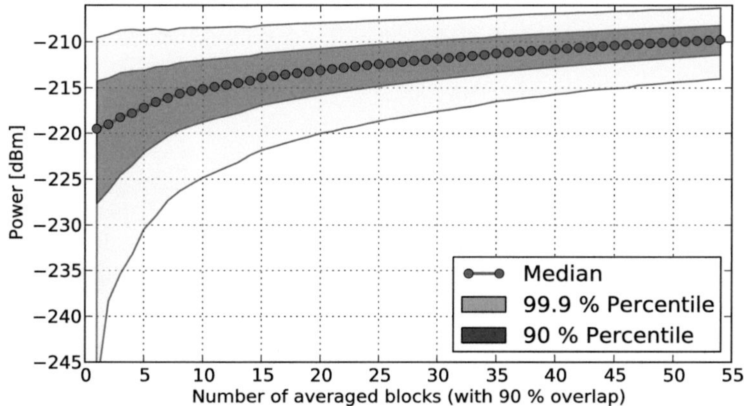

Figure 4.14.: In this example, a time trace of fixed length has been split into an increasing number of overlapping blocks (90 %). Each one has been processed by a DFT, averaging the results. This corresponds to Welch's method of spectral estimation. Note how the variance reduces with more averaged blocks. At the same time the average noise power does increase due to the smaller DFT blocksize. The overall effect is an increase of the 99.9 % Percentile of the background noise, indicating that averaging is not advantageous for signal detection.

Figure 4.14 shows the properties of the noise floor as a function of the averaged blocks for Welch's method. Averaging a larger number of smaller blocks increases the median noise power as expected. At the same time, the variance is reduced. Ultimately, the combined effect is not advantageous for signal detection, as indicated by the rising 99.9 % percentile.

In conclusion, best sensitivity is achieved by processing the entire time trace in a single DFT. Nonetheless, averaging several independent power spectra is useful if a time trace needs to be split into several (not continuous) parts to discard bad segments or due to limited acquisition memory.

4.4. Detection threshold

For the sake of clarity we will assume in the beginning of this chapter that the WISP signal will deposit all its power within a single frequency bin and the location of this bin is known in advance. We found that the latter assumption does not hold true in a practical measurement, because the frequency resolution of the RF sources (Anapico APSIN6010) is limited to 1 mHz. Due to rounding errors in the digital signal synthesis, the frequency of the output signal is not perfectly identical with the set point. In several experimental tests we have observed a **static** but unknown frequency offset on the order of 0.25 mHz. To consider this uncertainty in the location of the WISP signal, several frequency bins need to be tested for an excess in power. The statistical method which is first developed for testing a single bin, will later be adapted to the case of several bins.

The outcome of the experiment can be an exclusion result or a detected WISP candidate. To make a decision, we compare the power level of the frequency bin where the WISP signal is expected to a predefined detection threshold P_{th}. Finding a value for the detection threshold is an exercise in statistical hypothesis testing. The two possible outcomes of the experiment are:

$\mathbf{H_0}$: Null hypothesis. There is no WISP signal or its power is $< P_{sig}$. The frequency bin is governed exclusively by background noise.

$\mathbf{H_1}$: Alternative hypothesis. There is a WISP signal added to the background noise with a power of $\geq P_{sig}$.

We define the following variables for the hypothesis test:

$\mathbf{P}_{th}$ is the detection threshold. If the frequency bin where the WISP signal is expected, exceeds this power level, the outcome of the experiment will be positive.

$\mathbf{P}_{sig}$ is the minimum detectable signal power of the microwave receiver. In case of a negative result, this parameter is needed to derive the exclusion limit.

$\mathbf{PR}_\alpha := \mathbf{5\ \%}$ is the error of the first kind. This is the probability for a false positive outcome of the experiment. For example, H_0 is true and the WISP does not exist – however, the background noise exceeded the detection threshold, leading to a false detection.

$\mathbf{PR}_\beta := \mathbf{5\ \%}$ is the error of the second kind. This is the probability for a false negative outcome of the experiment. For example, H_1 is true, the WISP does exist and we receive a signal with an average power of $\geq P_{sig}$ – however, due to statistical fluctuations, the detection threshold was not exceeded.

If H_0 applies, the real and imaginary parts of the recorded baseband time trace will be distributed according to a Gaussian Probability Density Function (PDF) with a mean value of zero. The DFT operation (Eq. 4.2) will not change the PDF. Calculating the power spectrum involves taking the squared magnitude of the DFT result – which transforms the Gaussian PDF into a central χ^2 PDF with two degrees of freedom [91, 92]. If we average k subspectra, the degrees of freedom increase to $2k$. Taking this into account, the PDF for background noise in the power spectrum is:

$$\text{PDF}_{\mathrm{N}}(z) = \frac{z^{k-1}}{\sigma^{2k}\ 2^k\ \Gamma(k)}\ \exp\left(-\frac{z}{2\sigma^2}\right), \qquad \sigma^2 = \frac{P_N}{2k}, \qquad (4.8)$$

where $\Gamma(k)$ is the gamma function[2]. Note that the parameter σ^2 can be derived from the average noise power P_N of the spectral bins. When evaluating the measurement run, this parameter is estimated from the mean power of the frequency bins which are known to contain only background noise, allowing us to "fit" the analytical distribution to the experimental background noise. This has been demonstrated in Fig. 4.15, where the

---

[2]Defined as $\Gamma(z) = \int_0^\infty t^{z-1} e^{-t}\, \mathrm{d}t$.

histogram of 350000 frequency bins (shown as blue dots) from the ALPs run in June 2013 has been compared with the analytical PDF (shown as blue trace) from Eq. 4.8. The good agreement confirms, that the measured background noise is dominated by thermal noise from the detecting cavity, that there is no distortion due to the receiving chain or due to the signal processing and that the background noise is explained very well by the analytical PDF.

If the alternative hypothesis (H_1) applies, the recorded time trace consists of background noise and an additional sinusoidal signal, which could originate from reconverted WISPs. The real and imaginary part of the relevant frequency bin in the DFT-result will follow a Gaussian distribution with a non-zero mean value. In the power spectrum, the relevant bin will follow a non-central χ^2 distribution. The non-centrality parameter will be equal to the Root Mean Square (RMS) power of the sinusoidal signal P_{sig} [93, 94]. The distribution is given by:

$$\mathrm{PDF_{SN}}(z) = \frac{1}{2\sigma^2} \left(\frac{z}{P_{sig}} \right)^{\frac{k-1}{2}} \exp\left(-\frac{P_{sig} + z}{2\sigma^2} \right) I_{k-1}\left(\frac{\sqrt{P_{sig}\, z}}{\sigma^2} \right),$$

$$(4.9)$$

$$\sigma^2 = \frac{P_N}{2k},$$

where I_{k-1} is the modified Bessel function of order $k-1$.

To demonstrate that $\mathrm{PDF_{SN}}$ agrees with the statistics of a signal peak, a Monte Carlo simulation has been carried out. We computed 10000 power spectra of synthetic input data and computed the histogram of two specific frequency bins – one governed by noise and one with an additional signal. The corresponding analytical PDFs agree well with the histograms, as shown in Fig. 4.16.

The two error probabilities, PR_α and PR_β are related to the PDFs by:

$$PR_\alpha = \int_{P_{th}}^{\infty} \mathrm{PDF_N}(z)\, dz, \qquad (4.10)$$

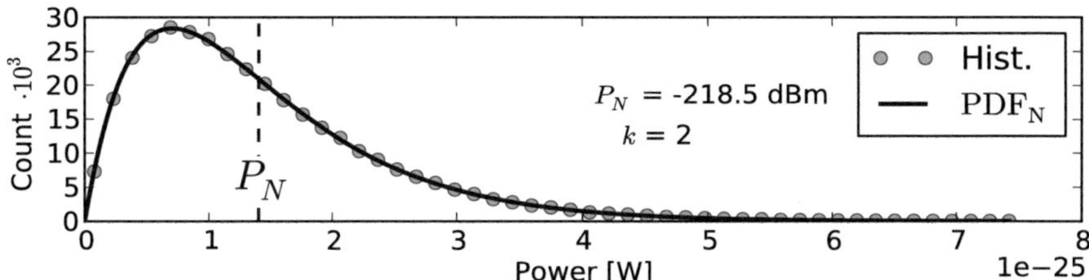

Figure 4.15.: Comparison of the analytical probability density function from Eq. 4.8 (PDF$_\mathrm{N}$) to the histogram of the background noise from the ALP measurement run in June 2013 (Hist.). The good agreement is indicating that the measured signal is dominated by thermal noise from the cavity and not distorted by the receiver or the signal processing steps.

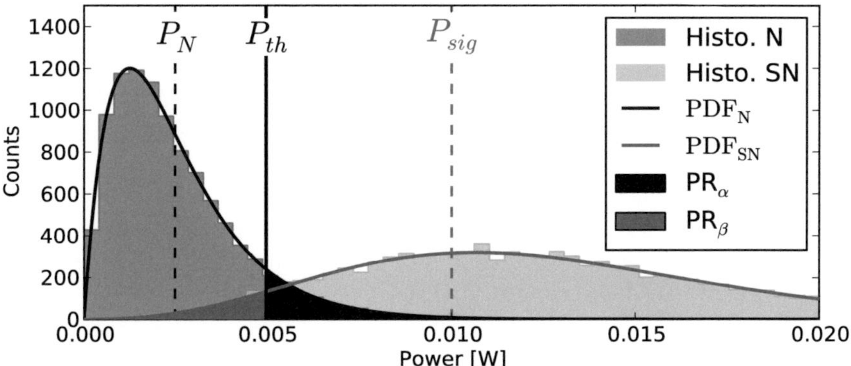

Figure 4.16.: Comparison of the analytical PDFs to a Monte Carlo simulation. The fluctuation of one spectral bin governed by background noise (N) and another bin with an additional sinusoidal signal (S+N) has been evaluated for 10000 synthetic power spectra. The analytical PDFs show good agreement with the histogram of the Monte Carlo data. Furthermore, an exemplary detection threshold P_{th} and the resulting error probabilities PR_α and PR_β have been indicated. Note that the power levels and probabilities are arbitrary and do not correspond to a measurement run.

$$PR_\beta = \int_0^{P_{th}} \mathrm{PDF}_{\mathrm{SN}}(z)\, dz. \tag{4.11}$$

This is illustrated by the dark blue and dark red areas in Fig. 4.16.

In order to define the detection threshold, the following steps were carried out after each measurement run:

1. The power spectrum was calculated. Approximately 300000 bins, which are known to exclusively contain background noise were selected. The average power of these bins was used to estimate P_N. The number of averaged subspectra (k) was known beforehand. Therefore the statistical properties of the background noise were completely defined.

2. With the known parameters from step 1 and with $\mathrm{PR}_\alpha := 5\ \%$, Eq. 4.10 was solved numerically for the detection threshold P_{th}.

3. With the definition of $\mathrm{PR}_\beta := 5\ \%$, the calculated P_{th} and the estimated P_N, Eq. 4.11 was solved numerically to obtain the minimum detectable signal power P_{sig}.

If the detection threshold was exceeded in a measurement run, we have detected a WISP candidate with a confidence level of $(1\text{-}\mathrm{PR}_\alpha) = 95\ \%$. If the detection threshold was not exceeded, we can state an exclusion limit from P_{sig} with a confidence level of $(1\text{-}PR_\beta) = 95\ \%$.

As mentioned in the beginning of this section, the exact location of the WISP frequency bin is not known. Therefore we search for a signal peak within a window of typically 400 μHz width. During the ALPs run in June 2012, 15 frequency bins fell into that window. This corresponds to carrying out the hypothesis test $T = 15$ times. In this case, the probability of getting a false positive can be expressed as a compound probability: it increases from $\mathrm{PR}_\alpha = 5\ \%$ to $\mathrm{PR}'_\alpha = 1 - (1 - \mathrm{PR}_\alpha)^T = 46\ \%$. To compensate for this effect, the false detection probability has been set to $\mathrm{PR}_\alpha := 5\ \%/T$, which results in a compound probability for getting a false positive of 4.9 % in the 15 bin case.

4.5. Phase stability

All free running microwave oscillators – therefore also the Anapico RF sources used in the experiment – suffer from small fluctuations of their output signals phase Φ and frequency $f = d\Phi/dt$. We distinguish between phase noise, which are fluctuations over short time spans (< 1 s) around a well defined mean value and frequency drifts, which happen over days and years, as the frequency determining element ages and permanently changes its properties. Phase noise is usually characterized in the frequency domain. The long term stability is usually specified as a fractional frequency shift $(\Delta f/f)$.

Both effects are problematic for the experiment's detection principle, as we effectively define an extremely narrowband filter around the WISP signal ($\mathrm{BW_{res}} \approx 10$ μHz). During the measurement, the frequency fluctuation of all oscillators must be $< \mathrm{BW_{res}}$. This involves the RF source driving the emitting cavity and any oscillator in the receiving chain. Otherwise, the WISP signal will misalign with the passband of the relevant frequency bin in the spectrum and cause a degradation of the detection sensitivity. The WISP signal will appear smeared out over several bins in the spectrum as demonstrated in an example measurement by the magenta trace in Fig. 4.17.

The worst case situation is met for the longest HSP measurement runs ($l \approx 30$ h) – in this case we require a fractional frequency accuracy of $\Delta f/f = 10$ μHz / 3 GHz $\approx 3 \cdot 10^{-15}$. The APSIN 6010 RF sources from the company Anapico are used in the experiment. When configured as independent and free-running oscillators, they are specified for a nominal frequency drift of $\Delta f/f = 5 \cdot 10^{-9}$ per day due to aging and $\pm 100 \cdot 10^{-9}$ over the temperature range of 0 °C … 50 °C. Both exceed the above requirements by far.

While frequency drifts of free running oscillators cannot be avoided, synchronizing them to a common reference clock can provide the necessary relative precision. The RF sources will still drift – but all of them in the same way. In a sense, the very narrowband filter for detecting a WISP signal will track and move with the slight frequency drift of the WISP signal. This has been achieved by means of a 10 MHz signal produced by the VSA and used as global frequency reference. An example of the

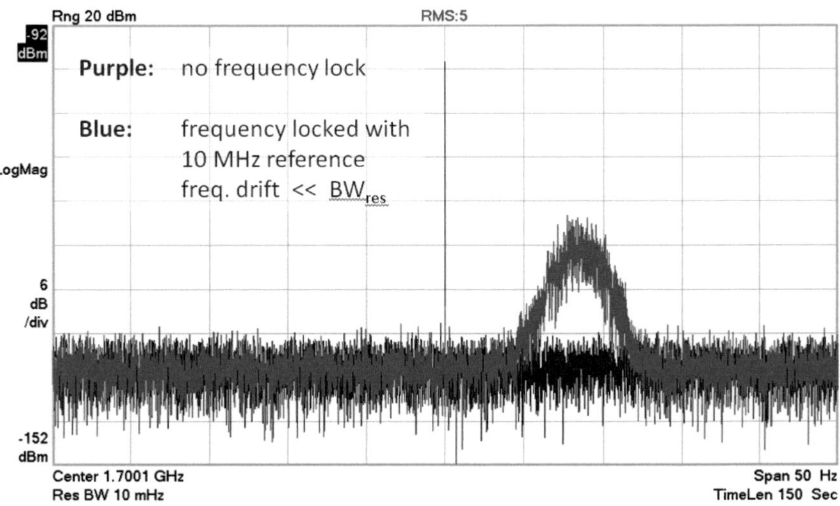

Figure 4.17.: Measurement demonstrating the importance of synchronizing all oscillators of the receiving chain to a common reference clock. In the synchronized case (blue), the test signal appears as clean signal peak at the expected frequency. Without synchronization (purple) the frequency drift and jitter of the two free-running oscillators smear out the signal peak. Moreover a static frequency offset of ≈ 7 Hz is visible due to the limited absolute frequency accuracy of the oscillators.

Figure 4.18.: Measurement setup for determining the relative phase drift between two Anapico RF sources at 2 GHz with a vector voltmeter. The 10 MHz reference signal has been used for phase locking.

improvement is given by the blue trace in Fig. 4.17, where a weak sinusoidal signal has been measured by the VSA with frequency synchronisation.

The effect of frequency drift is most critical for oscillators at higher frequency. Therefore the ANAPICO 6010 RF sources used for driving the emitting cavity and the first local oscillator in the custom made downmixing chain have been measured for their relative frequency precision. Two sources have been synchronized by their 10 MHz reference clocks. They were set to output RF signals with a frequency of $f_0 = 2$ GHz, which were fed into the ports of a HP 8508A vector voltmeter. This frequency was chosen as it corresponds to the upper limit of the vector voltmeter and as it falls between the frequencies used for ALP (1.7 GHz) and HSP (2.9 GHz) measurement runs. The instrument determines the phase difference between its input signals, which has been recorded over night to keep temperature fluctuations low.

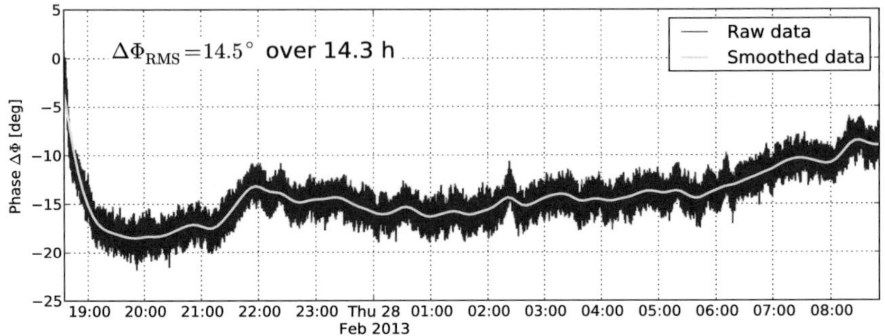

Figure 4.19.: Measured relative phase drift between two frequency locked Anapico RF sources at 2 GHz. The tolerable RMS phase drift for the narrowband signal detection is $\approx 180°$ over 14 h (see text).

Figure 4.19 indicates how – even though both frequencies were derived from a common reference clock – there is still no fixed phase relationship between the two output signals. The incoherency originates from the complex nature of the fractional Phase Locked Loop (PLL) within each RF source. The PLL basically multiplies the frequency of the reference signal by a fixed factor N to synthesize the output signal. Phase noise from the reference signal is amplified by $20 \cdot \log N$ and mirrored on the output signal, making the system proportionally more sensitive to phase noise and other distortions on the reference signal. Details on this are given in [95].

For the WISP measurement, phase deviations can be neglected, as long as the Root Mean Square (RMS) phase error $\Delta\Phi_{\mathrm{RMS}}$ is smaller than $360°$ over a timespan of $1/\mathrm{BW}_{\mathrm{res}} = 1/10\ \mu\mathrm{Hz} \approx 28$ h. The results from Fig. 4.19 indicate a deviation of $\Delta\Phi_{\mathrm{RMS}} \approx 15°$ over a 14 h time span, hence the synchronized RF sources are expected to be precise enough for the narrowband measurement and no broadening of the signal peak is expected. This has been confirmed by the power spectra of each experimental run, where a test tone signal is visible as narrow peak, with the minimum possible spectral width as dictated by the DFT. This can be verified, for example in Fig. 5.11, where the optimum response of the DFT has been overlaid with the spectral results.

For precise frequency standards it is common practice to quote the "Allan variance" (σ_a). This is a statistical tool to describe the long term stability of an oscillator in the time domain. The frequency fluctuation is measured and averaged over a number of different integration times, which allows us to resolve the oscillators characteristics over different time scales.

Computing this parameter for the phase-locked RF sources allows us to compare their long term stability to other oscillators and hence gain some interesting insights on the achieved performance. It is possible to convert the measured data from Fig. 4.19 into an Allan variance. First, the measured phase error needs to be converted into a instantaneous frequency $(f(t))$ and then into a relative frequency error $(f_{\rm rel})$ by:

$$f(t) = \frac{d\Delta\Phi}{dt} + f_0, \qquad f_{\rm rel} = \frac{f(t)}{f_0} \tag{4.12}$$

where $f_0 = 2$ GHz is the measurement frequency. As a next step, $f_{\rm rel}$ is divided into a number of M segments, each one with a length equal to the integration time τ. For each segment, with the index k, the mean frequency error (f_k) is calculated:

$$f_k = \text{mean}\left(f_{\rm rel}\right), \qquad \text{with } k = 1 \ldots M. \tag{4.13}$$

With these suppositions, the Allan variance is given by the difference of neighboring frequency errors:

$$\sigma_a^2(\tau) = \frac{1}{2(M-1)} \sum_{k=1}^{M-1} \left(f_{k+1} - f_k\right)^2. \tag{4.14}$$

The measured Allan variance of the two phase locked RF sources, as a function of integration time τ, is shown in Fig. 4.20. Compared to the RMS variance, the Allan variance has the advantage of only being sensitive to frequency drifts happening on a time-scale defined by the integration time τ. Faster frequency drifts on a timescale $\ll \tau$ are suppressed by the averaging operation. Slower frequency drifts on a timescale $\gg \tau$ are suppressed by the fact, that only the difference in frequency error between

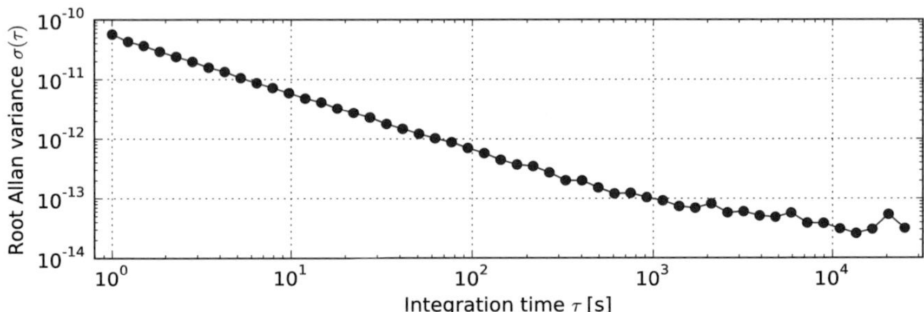

Figure 4.20.: Relative Allan variance (σ_a) between the two RF sources, phase locked by a 10 MHz signal. The data points above $\tau = 10^4$ s are not well defined due to the limited measurement time of 14 h.

two blocks is taken into account. These properties make it possible to distinguish between drift processes happening on different time scales.

As the Allan variance is commonly quoted in the datasheet of frequency standards, the result from Fig. 4.20 can be compared to the stability of other precision oscillators. Some of the most accurate and stable clocks are used in the NAVSTAR satellites, which operate the Global Positioning System (GPS). They are based on Rubidium and Caesium standards and as their accuracy is critical for the functioning of GPS, their parameters have been studied in detail [96]. With an integration time of $\tau = 10^4$ s they typically achieve an Allan variance of $\sigma_a = 10^{-12}$ and are therefore less stable than the two phase locked RF sources. However it must be kept in mind that we have measured relative frequency offsets. The NAVSTAR satellites have to provide absolute frequency accuracy, which is much harder to achieve.

4.6. Design and realization of a heterodyne receiving chain

For all the ALP and HSP measurement runs up to the time of writing, a commercial Vector Signal Analyzer (VSA) was used to acquire the microwave signal from the front-end. We were motivated to replace this device with a custom built downmixing chain for several reasons:

- For a typical 29 h measurement run, the VSA cannot record a wider span than $SP = 2$ kHz due to its limited acquisition memory (see Eq. 4.1). An increased span would allow us to see the resonance curve of the detecting cavity in the spectrum. Moreover, we would have a better time resolution to resolve transients, interference or unexpected signals in the time domain.

- At the beginning of the CROWS experiment, it was not clear if the VSA can achieve the required frequency and phase stability. Furthermore, the VSA (and especially its software suite) is a general purpose instrument, not designed for recording and evaluating data over several hours.

- It would allow us more control over each signal processing step than the commercial VSA, which had to be used as a black box.

- It would have the ability to track frequency swept RF sources, which would allow us to exploit accelerator cavities as parasitic hidden photon emitters. The frequency tracking would enable us to process the data by the narrowband receiving concept.

- It would provide a cost advantage by a factor of ≈ 10 compared to the Agilent VSA, which might be interesting for future LSW experiments.

The custom downmixing chain needs to be able to isolate and record a slice in the spectrum with up to 20 kHz span, centered at an arbitrary frequency in the range of 1 - 3 GHz. Larger spans would lead to too much recorded data over a 29 h run. The recording time should be indefinite and only limited by the available disk space on the data acquisition PC. All oscillators involved in the frequency and analog to digital conversion process must be synchronized to a common 10 MHz reference – this is a

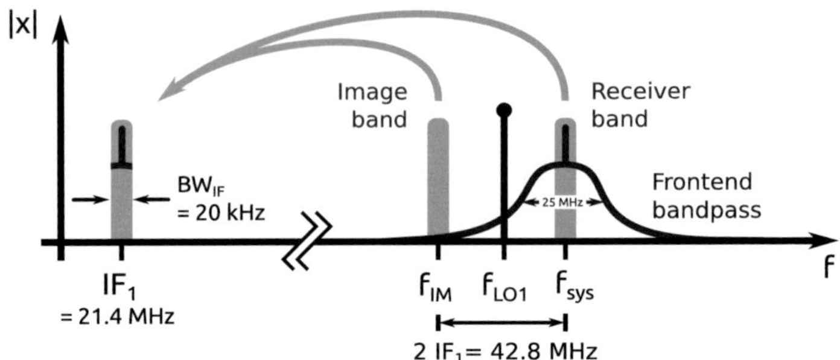

Figure 4.21.: Illustration of the receiver and image band for the first frequency conversion.

requirement of the very narrowband signal detection concept. Moreover, the chain should be able to provide up to 80 dB of gain (adjustable and spread over several intermediate frequencies).

The design choice fell on a double heterodyne receiver concept. The architecture makes use of two frequency conversion steps. The first one converts the microwave signal to an intermediate frequency (IF) in the MHz range. In a second step, the signal is converted from the IF to a lower frequency of several kHz, which can be easily handled by a high-resolution Analog to Digital Converter (ADC). All further signal processing is then carried out in the digital domain.

The double heterodyne architecture has two distinct advantages:

Selectivity: The first intermediate frequency (IF$_1$) was chosen in the MHz range, allowing us to exploit the advantages of very sharp and high performance crystal filters, giving the receiver excellent selectivity (and hence rendering it immune against out of band interference). The measured IF filter response is shown in Fig. 4.23.

Image rejection: Due to an ambiguity in the mixing process, a heterodyne architecture will be sensitive to two frequency bands. One is the wanted "receiver band" (centered at f_{sys}), the other is the unwanted image band (centered at $f_{IM} = f_{sys} - 2\,IF_1$). This has been illustrated in Fig. 4.21. We are motivated to keep a large distance between both bands, making it easier to block the image band by a filter on the

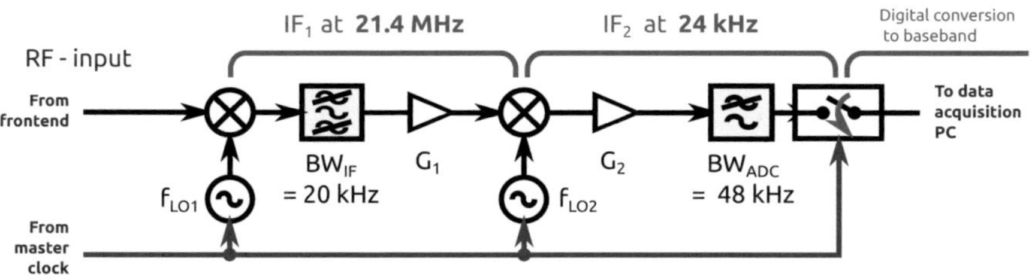

Figure 4.22.: Block diagram of the downmixing chain, showing the two analog frequency conversion stages and the relevant components involved.

input of the receiver. The double heterodyne architecture allows us to choose the first intermediate frequency relatively high – in our case IF$_1$ = 21.4 MHz, so the image band will fall 42.8 MHz below the receiver band. This simplifies the design as the front-end filter (described in 3.5.2) with a bandwidth of 25 MHz provides already enough image rejection and no additional filter is necessary.

A schematic of the downmixing chain is shown in Fig. 4.22. We will now look into the function of each component and the signals at each point: The input to the mixing chain is a noise-like signal from the front-end, which is centered at $f_\mathrm{sys} \approx 2.9$ GHz (for ALP search). Note that only a single cavity mode falls into the passband of the front-end filter, all others are suppressed and cannot interfere with signal detection. The first mixer transforms the signal from 2.9 GHz to the first Intermediate Frequency IF$_1$ = 21.4 MHz. The local oscillator signal (LO$_1$) is provided by an Anapico RF source, which is set to the frequency $f_{LO1} = f_\mathrm{sys} - 21.4$ MHz. Note that the center frequency of the receiver chain can be easily adjusted over a wide range by changing f_{LO1}. After the mixer, the IF signal is filtered by a crystal based bandpass filter with a bandwidth of BW$_{IF}$ = 20 kHz. It is this component, which determines the final "IF bandwidth" of the receiver, equivalent to the maximum span we can see in a recorded spectrum. The filter (Quartzcom 21Q20D-145) has 8 poles, resulting in a passband with a very steep cut-off (a measurement is shown in Fig. 4.23).

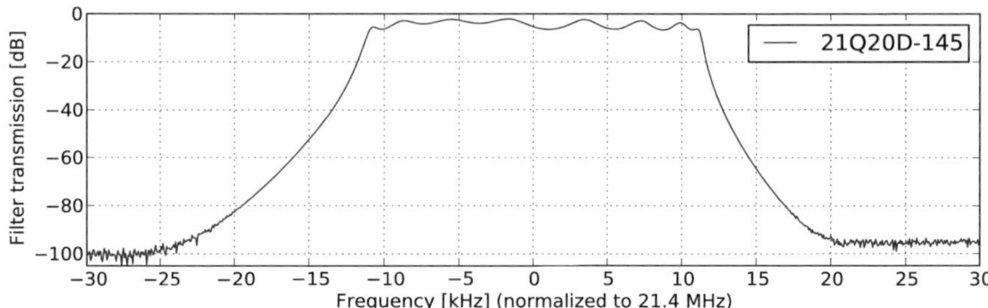

Figure 4.23.: Measurement of the IF filter (Quartzcom 21Q20D-145) passband in transmission. The frequency response of the receiver is determined by this filter.

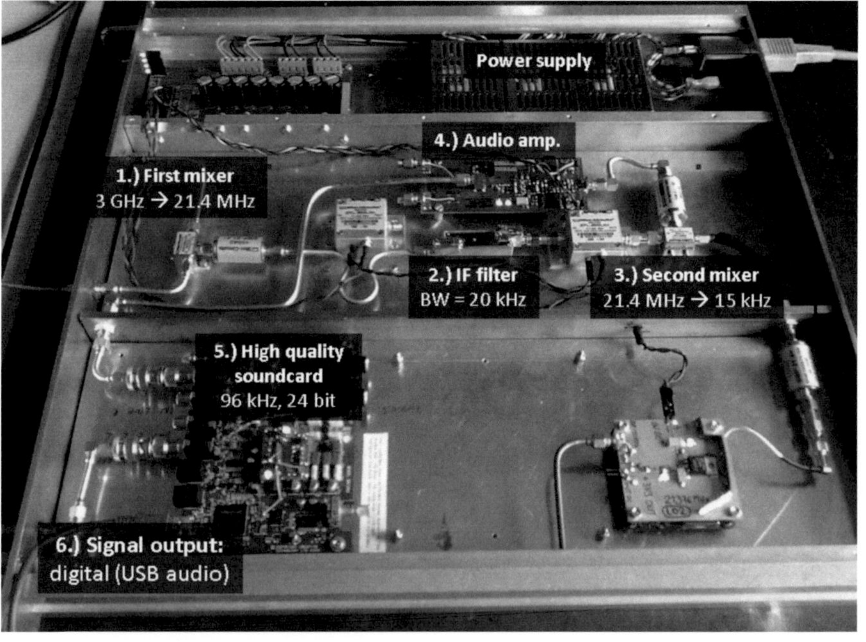

Figure 4.24.: Photo of the custom downmixing chain, as it has been used to record data during the ALP measurement run in June 2013.

4.6.1. Analog to digital conversion

The spectral content after the IF filter is limited to a 20 kHz wide band centered at 21.4 MHz. Our goal is to sample this signal and convert it to the digital domain.

There are several options to achieve this. Direct conversion by a fast Analog to Digital Converter (ADC) is possible. As the signal is band-limited, the sampling rate of the ADC can be lower than the Nyquist rate (in fact, theoretically as low as 50 kHz). In this case, careful control of aliasing products is necessary, especially for the large undersampling ratio involved. Another option is to convert the IF signal to a baseband signal with an analog IQ mixer in hardware. The resulting quadrature signal requires two ADCs (for I and Q respectively) with a sampling rate > 40 kHz. This solution was not chosen as the analog hardware introduces distortions – the I and Q signal paths are never perfectly symmetric and need to be carefully calibrated (in hardware or software). Moreover, the analog IQ mixer usually suffers from a blind spot at 0 Hz in the resulting two sided spectrum, corresponding to f_{sys}, which is the most important frequency in our case.

We avoid these disadvantages by using a second frequency translation stage, shifting the signal to a second intermediate frequency of $IF_2 = 24$ kHz – a frequency which can be handled by high quality audio ADCs. The second mixer is driven by a LO with a fixed frequency of $f_{LO2} = 21.4$ MHz-24 kHz. The signal on the mixers output is amplified, low pass filtered to prevent aliasing and fed into the ADC.

We used an ADC with 96 kHz sampling rate and 24 bit resolution (Cirrus Logic CS5361). The usable frequency band of the ADC extends from 0.1 kHz to 48 kHz, from which only the range of 14 kHz up to 34 kHz is occupied by the input signal. This provides a wide "guard band", to accommodate the transition region of the IF filter and hence prevent aliasing. In a future upgrade, the IF filter might be replaced to allow for a wider bandwidth. The digitized data is transferred from the ADC to the host PC over an Universal Serial Bus (USB) interface.

Instead of implementing the ADC and USB interface from scratch, a commercial USB soundcard (Creative Sound Blaster X-Fi HD) was modified for our purposes. The device is based on the aforementioned high performance ADC chip. The idea was to save time and effort on the hard and software development, as the downmixed data-stream could be transferred over the standardized audio drivers of the Windows or Linux operating system. The soundcard was modified to make it compatible with the narrowband detection concept. A Phase Locked Loop (PLL) circuit was used to synchronize the devices system clock (24.576 MHz) with an external 10 MHz reference signal.

We identified two options to continue the signal processing on the host PC side. The most straightforward way is to store the raw recorded data directly to disk, using a standard audio recording program and do all the further data processing offline. The biggest disadvantage of this approach is the lack of flexibility. Data has to be recorded to disk with a sampling rate of 96 kHz, even though the usable information is conveyed in a only 20 kHz wide band, centered at IF_2. While the available disk space is in general not a bottleneck with modern computers, the superfluous data generally complicates the offline processing and evaluation. Moreover, one might think of a situation where only a narrow slice in the spectrum is of interest. It would be advantageous, if only the information within this spectral slice could be recorded to disk.

A more versatile and slightly more complex approach is to use the software suite "GNU radio" for the final frequency conversion and recording step. The open source software allows realtime processing of digital data streams in a user friendly way by specifying a signal flow-graph. A flow-graph has been developed, which converts a user definable center frequency (f_c) and span to baseband and records the filtered and decimated data to disk, it is shown in Fig. 4.25. The first signal processing step is a multiplication with a complex sinusoid of frequency $-f_c$. This corresponds to a digital quadrature demodulation. The user defined center frequency (f_c) is shifted to baseband, resulting in a complex signal with In-phase (I) and Quadrature (Q) components and a two sided spectrum. A digital lowpass filter band-limits the signal to the selected span. Then decimation is applied, reducing the sampling rate to the minimum possible value while avoiding information loss due to aliasing. The resulting filtered and decimated baseband signal is streamed to a file on disk for further evaluation. Furthermore, its waveform and spectrum can be visualized in realtime, which is useful for development and monitoring purposes.

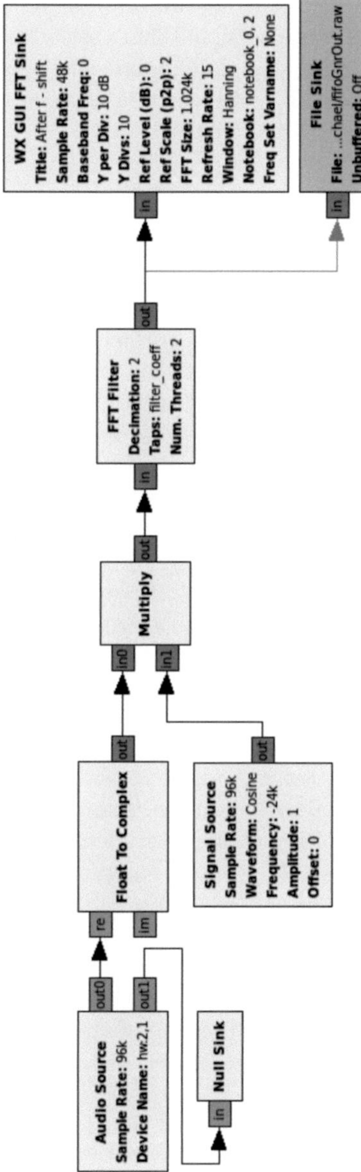

Figure 4.25.: GNU radio flow graph, converting the digitized signal from $IF_2 = 24$ kHz to a quadrature baseband signal and storing it to disk for further evaluation. The center-frequency and decimation factor is user configurable.

The GNU Radio approach renders the downmixing chain more powerful since it can be used as general purpose signal analyzer with user-definable center frequency (by adjusting f_{LO1}) and span. In contrast to most commercial instruments, it is capable of recording continuous time traces, only limited by the available hard-disk space.

4.6.2. Lost Samples

The narrowband signal detection concept, as described in 4.2 requires that all samples of the time domain trace are acquired in a continuous fashion with a constant amount of time between each sample. As we are searching for a sinusoidal signal, the loss of one or several samples would appear like a phase jump in the recorded time trace. In the frequency domain, the response function of this phase jump would be folded into the spectrum and smear out the signal peak which we are trying to detect with a very narrowband filter. As the signal power will be spread over several frequency bins, this would reduce the signal to noise ratio in the bin which we are testing for a WISP signal. In the case of an exclusion limit, no signal peak is visible but the lost samples would introduce an error in the estimation of the minimum detectable signal power.

Hence we require a continuous data acquisition which does not drop a single sample throughout the entire recording time. For the commercial VSA, this is guaranteed by the manufacturer and made possible by recording data into a dedicated acquisition memory on hardware level. One disadvantage of this concept is, that the size of the acquisition memory is finite, limiting the maximum number of continuously acquired samples to $N_{max} = 268 \cdot 10^6$. For the custom downmixing chain, there is no such limitation: data is streamed over USB to a PC, where it is stored on a harddisk, providing several orders of magnitude more storage space than the aforementioned acquisition memory. On the other hand, it critically depends on the realtime capabilities and the driver architecture of the operating system, if the data stream can be processed and stored to disk reliably and without any lost samples.

Two operating systems have been tested for their capability to record a continuous time trace. The two tested configurations were:

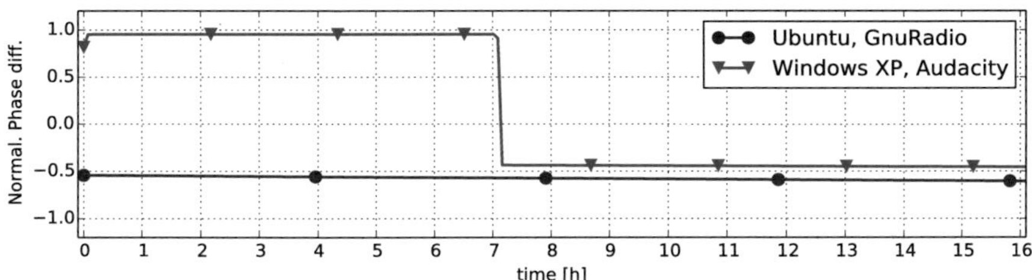

Figure 4.26.: Test for lost samples. Any lost samples would be indicated by sudden jumps of the trace. For the Windows based recording, samples were lost at t ≈ 7 h.

1. Ubuntu 12.04 LTS (a Linux distribution) with the "GNU Radio 3.6.3" flow-graph, as depicted in Fig. 4.25

2. Windows XP with the audio editor "Audacity 2.0.5"

For the test, a signal generator (HP 33120A) was used to generate a sine-wave with an arbitrary frequency of $f_{test} = 3.456$ kHz. The sampling clock of the ADC was synchronized to the signal generator by means of a 10 MHz reference signal. The test signal was recorded for 16 h with a sampling rate of 96 kHz under both configurations.

To evaluate the recorded time trace for discontinuities, the relative phase shift (φ_{meas}) of the recorded sine wave to a synthetic "reference" sine wave of identical frequency was measured. The phase shift can be determined by multiplying both signals, as will be shown. The recorded signal is given by $A_1 \sin(\omega_1 t + \varphi_1)$, where A_1 is its amplitude, ω_1 its frequency and φ_1 its phase. It is multiplied with the synthetic sinewave with the parameters A_2, $\omega_2 = \omega_1$ and φ_2. The results of this multiplication are:

$$\varphi_{meas} = A_1 \sin(\omega_1 t + \varphi_1) \cdot A_2 \sin(\omega_2 t + \varphi_2) \qquad (4.15)$$

$$= \frac{1}{2} A_1 A_2 \left[\cos\left((\omega_1 - \omega_2)t + \varphi_1 - \varphi_2\right) - \cos\left((\omega_1 + \omega_2)t + \varphi_1 + \varphi_2\right) \right].$$

Note how the resulting signal has a term at the sum and difference frequency. The former one is eliminated by a lowpass filter with very low cut-off

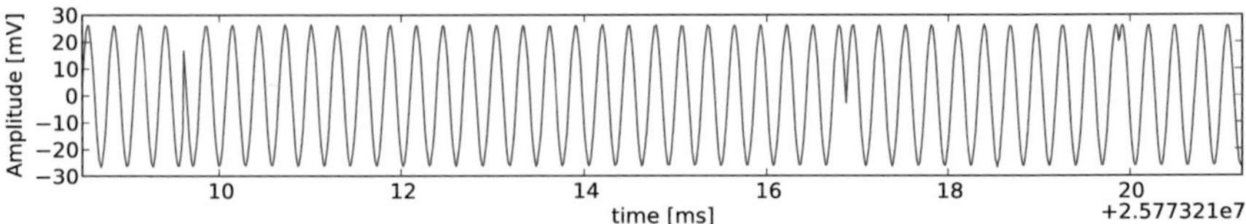

Figure 4.27.: Unprocessed time domain trace of the Windows recording, zoomed on t = 7 h, showing that several samples were lost, probably due to the limited realtime capability of the operating system.

frequency. The latter one is zero, as ω_1 and ω_2 are of the same frequency. Taking this into account, the equation simplifies to:

$$\varphi_{meas} = \frac{1}{2} A_1 A_2 \cos(\varphi_1 - \varphi_2). \tag{4.16}$$

Note that the amplitude dependence has been removed by normalizing to $-1 \leq \varphi_{meas} \leq 1$. The resulting plot is shown in Fig. 4.26. Any lost samples would lead to a sudden jump of the trace. We will now determine how large this jump is for a single lost sample. At a sampling rate of $f_s = 96$ kHz, losing a single sample would correspond to skipping forward in time by $1/f_s = 10.4\ \mu$s. This corresponds to a phase jump of $\Delta\varphi_e = 0.226$ rad $= 13°$ for the test signal at $f_{test} = 3.456$ kHz. The phase error for one lost sample is hence given by:

$$\Delta\varphi_e = \frac{f_{test}}{f_s} 2\pi. \tag{4.17}$$

Note that we do not plot the phase error directly, but the normalized $\cos()$ function of it, hence the trace would jump by $\Delta\varphi_{meas} \approx \Delta\varphi_e/\pi = 0.07$, which should be clearly visible in the figure.

The results in Fig. 4.26 indicate that the recording under Linux with GNU Radio did not drop a single sample over 16 h – hence this configuration is usable for experimental recordings. On the other hand, the configuration under Windows XP with Audacity lost several samples at t $\approx$ 7 h. The unprocessed time trace, zoomed over the relevant range in time is shown in Fig. 4.27 and indicates that an unknown number of samples were lost in

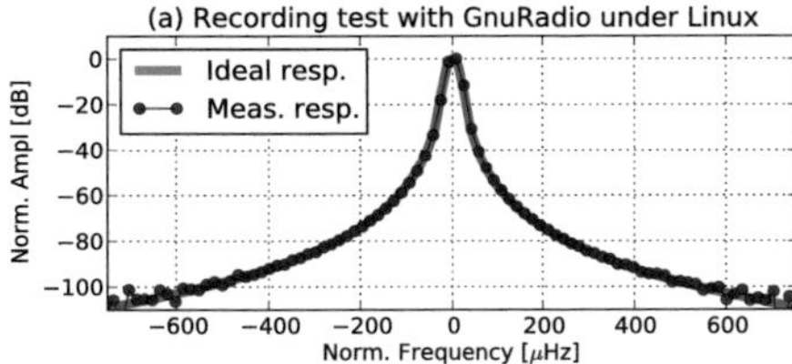
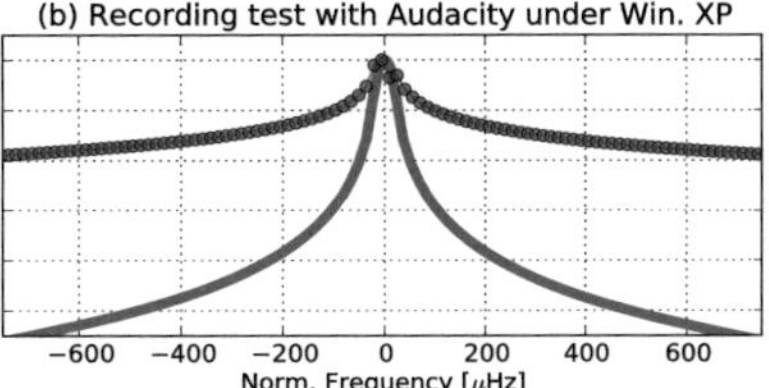

Figure 4.28.: Evaluation for lost samples in the frequency domain. No samples were lost in (a) and the measured signal peak (blue) does agree with the frequency response of the windowing function (grey). Samples were lost in (b), which leads to a smeared out signal peak and degraded signal to noise ratio. Both plots share the same Y-axis.

three bursts within a 20 ms period. One should note that the tested Linux operating system provides special low-latency and realtime scheduling options, which can ensure that the data from the soundcard is processed by the sound driver within a short time of acquisition, preventing buffer overflows. Windows XP does not provide these features and hence is a less reliable option to record the data stream – buffer overflows and lost samples are more likely.

To demonstrate the impact of the lost samples on the narrowband signal detection concept, the recorded data has been evaluated in the frequency domain. The shape of an ideal signal peak, without any lost samples, has been illustrated by a grey trace. It falls off at the maximum possible rate, as dictated by the Hanning window (its frequency response has been shown in Fig. 4.10). The GNU Radio recording (blue) in Fig. 4.28(a) agrees very well with this ideal peak. The Windows recording (green) in Fig. 4.28(b) demonstrates how even a small number of lost samples will smear out the peak, resulting in spectral leakage and a broader effective resolution bandwidth, which will degrade the detection sensitivity in an experimental run.

In conclusion, valid experimental results can be derived from the custom downmixing chain if Linux and the GNU Radio software are used for recording.

In June 2013, the ALP measurement run was carried out with both receivers, the VSA and the custom downmixing chain, recording data in parallel. Unfortunately, the custom downmixing chain was used in combination with Windows XP and Audacity. The tendency of this configuration to drop samples was not known at the time. No dedicated data acquisition PC was available and the space and power dissipation within the receiver shielding enclosure were limited. These problems forced us to use the embedded control PC within the VSA for data acquisition, which limited us to the use of Windows.

At first glance, the time trace was recorded successfully by the custom downmixing chain with the Audacity software. Nonetheless, the benchmark results from this chapter make it very likely that several samples have been dropped throughout the recording, making the data from the custom downmixing chain unsuitable for an exclusion limit calculation by the narrowband detection concept. Note that the exclusion limits in 5.1 were derived from the data recorded by the VSA hardware, which is guaranteed to not drop any samples under any conditions.

4.6.3. Possible data evaluation

The recorded baseband signal from the custom downmixing chain can be processed in the same way as described in 4.2 – provided it has been acquired without any dropped samples. Moreover, the wider bandwidth of the recording, compared to the VSA data, allows us to identify transient or time dependent interference signals, which would indicate either a problem with the electromagnetic shielding or could be of potential physical origin and should be further investigated. For that reason, the spectrogram of the recorded data has been determined. It corresponds to a low resolution spectrum, where the amplitude values have been color coded and plotted over time. In this analysis, the time trace is split into many sections and processed by smaller DFTs, hence lost samples do not influence the result significantly.

One example of such a resulting spectrogram is shown in Fig. 4.29. The data has been recorded during the ALP measurement run in June 2013 with the custom downmixing chain. The sampling rate was set to 96 kHz and no decimation or filtering was applied, hence the plots show the full

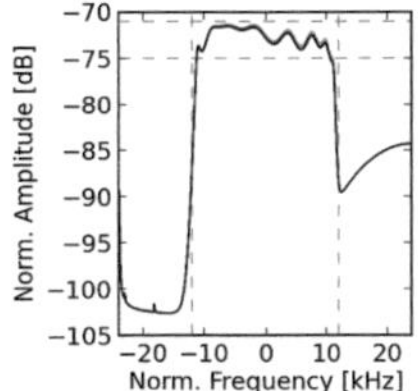

(a) Low resolution spectrum, showing the IF-filter passband.

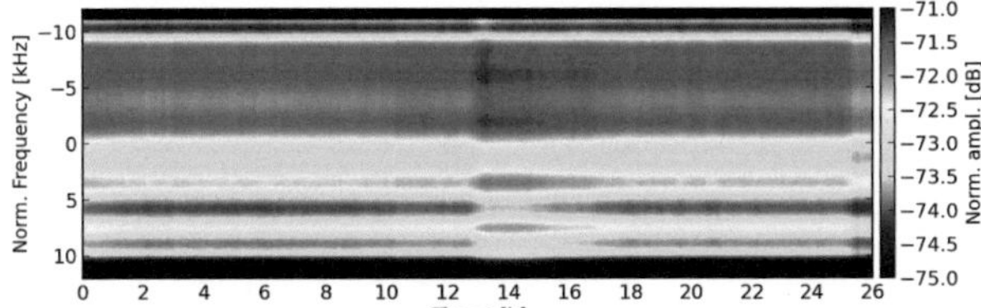

(b) Low resolution spectrogram. No spurious signals are visible. The sudden increase of noise power at t=12 h is caused by a temperature fluctuation, changing the detecting cavities resonant frequency (see also 5.1.2).

Figure 4.29.: Recorded data of the custom made downmixing chain during the ALP measurement run in June 2013. The passband ripple of the IF filter leads to a strong structure in the noise floor. Note that the amplitudes have been normalized to the ADC range. No calibration for absolute power has been applied in this measurement.

20 kHz wide response of the IF filter and the guard band above and below it, which is best visible in the spectrum Fig. 4.29a. The spectrogram in Fig. 4.29b shows the variation of the spectrum over time and hence is a useful tool to identify time dependent interferes or unexpected signals. It has been normalized to the frequency and amplitude range, indicated in Fig. 4.29a by the dashed green lines. While the vertical structures in the spectrogram originate from the ripples of the IF filter and are hence an artifact of the downmixing chain, there is a transient visible at t = 13 h. This sudden increase in noise power could be traced back to a temperature fluctuation, which changed the resonant frequency of the detecting cavity and hence the received noise power. Further details on this are given in 5.1.2. Besides this fluctuation of known origin, the spectrogram does not show any other unexpected signals within its 20 kHz bandwidth and hence demonstrates that no interference could penetrate the electromagnetic shielding and perturb the measurement results in this band.

4.7. Very high resolution spectral estimation

During an experimental run, the length of a typical time trace is 30 h. If data is recorded by the VSA, the typical span is SP = 2.5 kHz. With the custom downmixing chain up to SP = 20 kHz are possible. A recording will hence consist of up to $2 \cdot 10^9$ complex samples. They are stored in the form of two floating point numbers, represented by 8 bytes each. The file size of the recorded data is hence several tens of GB. Considering the additional zero padding of the recorded data (as described in 4.3.2), file size increases to several hundreds of GB.

As described in 4.2, the signal processing requires us to apply a single DFT operation over the entire recorded time trace. One approach is to make use of the software library FFTW [97], which can process DFTs of arbitrary size and is heavily optimized for speed. For example, during a test, a DFT operation of 10^8 samples has been carried out in typically 5 s but required more than 3 GB of Random Access Memory (RAM) on a PC, which was up to date at the time of writing[3]. While time is not a problem, the required amount of RAM scales with DFT size. Larger DFTs in the order of 10^{10} will require specialized servers or supercomputers with large amount of RAM.

4.7.1. Splitting the DFT operation

A solution to avoid this RAM bottleneck and to process DFTs of arbitrary size on an average PC was found: It is possible to split one big DFT operation over N samples into $2\sqrt{N}$ operations over $\sqrt{N}$ samples by the so called "Four step" algorithm [98, 99]. This allows the smaller DFT operations to be carried out in memory (in-core), while the larger part of the data is stored on an external disk (out-of-core). With this algorithm, the size of the DFT operation will only be limited by the amount of available hard disk space, which is in general several orders of magnitude larger than the available RAM.

The algorithm works as follows (the steps have been illustrated in Fig. 4.30):

---

[3]Intel© Core i7-2640M CPU at 2.80 GHz × 4 with 16 GB of RAM

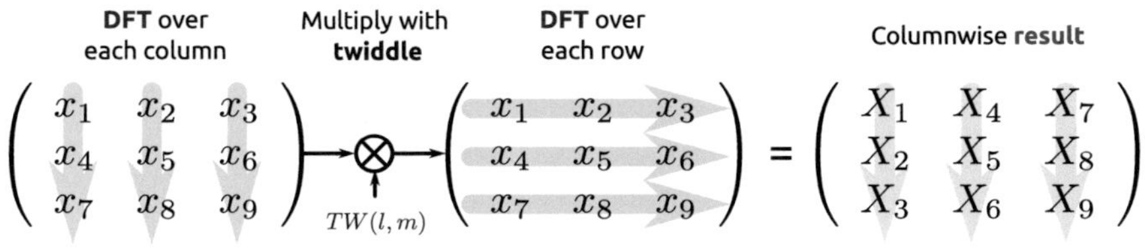

$$
\underbrace{\begin{pmatrix} x_1 & x_2 & x_3 \\ x_4 & x_5 & x_6 \\ x_7 & x_8 & x_9 \end{pmatrix}}_{\text{DFT over each column}} \longrightarrow \underset{TW(l,m)}{\otimes} \longrightarrow \underbrace{\begin{pmatrix} x_1 & x_2 & x_3 \\ x_4 & x_5 & x_6 \\ x_7 & x_8 & x_9 \end{pmatrix}}_{\text{DFT over each row}} = \underbrace{\begin{pmatrix} X_1 & X_4 & X_7 \\ X_2 & X_5 & X_8 \\ X_3 & X_6 & X_9 \end{pmatrix}}_{\text{Columnwise result}}
$$

Figure 4.30.: Illustration of the 4-step algorithm. The input time series $(x_1 \ldots x_9)$ is transformed into the frequency domain $(X_1 \ldots X_9)$ by 6 small DFT operations.

1. The N samples of the time trace are arranged into a $\sqrt{N} \times \sqrt{N}$ matrix, where subsequent samples are stored row-wise. Then a DFT operation is carried out for each column in the matrix.

2. Each element in the matrix is multiplied by a "twiddle factor", as defined by Eq. 4.18.

3. The DFT operation is carried out for each row in the matrix.

4. The result of the DFT over the whole dataset can be read column-wise from the matrix.

The twiddle factors correspond to root-of-unity multiplicative constants for each matrix element and are defined as:

$$
TW(l, m) = \exp\left(-j2\pi lm/N\right), \tag{4.18}
$$

where l is the row and m the column index.

This algorithm allows us to calculate arbitrarily large DFTs, limited only by the amount of available disk space. It trades speed for RAM usage. Nonetheless, one challenge remains: For each step, data is read from disk, processed and wrote back. As we need the data organized in a 2D matrix, which can be accessed row and column-wise, a special file format is necessary, as explained in 4.7.2.

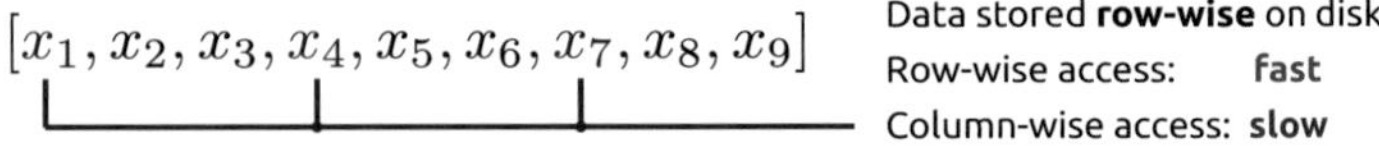

Figure 4.31.: If data is stored row-wise on disk, column-wise access is slow
due to the large number of seek operations.

4.7.2. Transposing a large disk stored matrix efficiently

The previously described "4-Step" algorithm requires the recorded time-
trace to be interpreted in a matrix format. On the other hand, the
data is stored on hard-disk in a linear fashion. Accessing the data row-
wise $(x_1, x_2, x_3, \ldots)$ is fast, as it corresponds to the arrangement on the
disk – however accessing the data column-wise $(x_1, x_4, x_7, \ldots)$ is extremely
slow, as the hard-disk needs to seek over $\sqrt{N}$ samples for each single read
operation, as shown in Fig. 4.31. The same would be true for a column-wise
file which is accessed in a row-wise fashion.

A solution to this bottleneck is to store the data on disk in a more complex,
intermediate format, which is neither column or row-wise. By arranging
the samples into sub-matrices in an advantageous way it is possible to read
the data row or column wise with similar performance from disk. This
kind of intermediate format has been described in [100] and shall not be
explained here in detail.

A software library to read and write large data sets in a row- or column-
wise fashion from disk (according to [100]) has been implemented in the
programming language Python. This library was used to realize very large
DFTs by the 4-step algorithm, as described in 4.7.2. The largest single
DFT operation using this principle was carried out over a time-series
consisting of $2 \cdot 10^{10}$ samples, which took ≈ 5 h on the aforementioned
state-of-the-art PC[1]. An external USB harddisk was used for data storage,
which severely limited the speed of the operation – however, as time is
not critical in the offline data analysis, no speed optimizations have been
applied so far.

The numerical accuracy (and correctness) of the implementation of the
4-step algorithm, was tested by comparing it to the FFTW library, which
was known to provide results with a well-defined accuracy. For this

purpose, an array of 10^8 random complex numbers were transformed by both algorithms and the two results ($X_{4\text{step}}$ and X_{FFTW}) were compared by the following metric:

$$A_{\text{RMS}} = \sqrt{\sum_N \left| \frac{X_{4\text{step}} - X_{\text{FFTW}}}{X_{\text{FFTW}}} \right|^2}, \qquad (4.19)$$

it corresponds to the relative RMS error between the two results. For the 10^8 points DFT, we measured $A_{\text{RMS}} = 5.918 \cdot 10^{-16}$. This proves that the results of the 4-step DFT agree well with the results of the FFTW library. Note that 64 bit floating point numbers have been used for the calculation. Their maximum resolution is limited to $1.1 \cdot 10^{-16}$, which is close to the achieved RMS error.

5. Achieved exclusion results

Within the 3 years timespan of this work, the experimental setup has been continuously improved. An overview of all measurement runs with their most important technical parameters is given in Table 5.1.

The most visible changes have been applied to the EM shielding enclosure for the detecting cavity. For the earliest HSP run, no dedicated enclosure was used, instead it was placed within the VSA shielding box, as shown in Fig. 5.1(a). Subsequent HSP runs until 12/2012 featured a recuperated stainless steel shielding enclosure for the detecting cavity, depicted in Fig. 5.1(b). It allowed first ALP measurements in a 0.5 T magnet and served as a prototype for the final aluminium enclosure, shown in Fig. 5.1(c) and described in 3.4.3. It has been utilized in the last two – and most sensitive – measurement runs for ALPs and HSPs. Apart from the magnet, both runs were carried out with an identical apparatus. The recorded data from these runs has been evaluated to produce the most stringent exclusion limits. The derivation of these results will be described in detail in the subsequent sections.

The flow graph in Fig. 5.2 shows an overview of the typical steps of the experimental procedure.

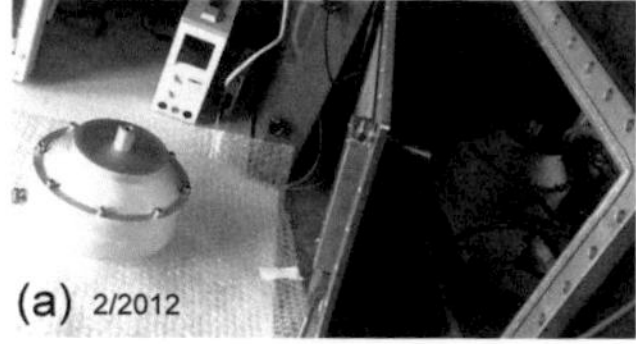

Figure 5.1.: Evolution of the shielding enclosure for the detecting cavity.

Table 5.1.: Overview of all measurement runs for HSPs and ALPs. $\mathbf{P}_{\mathrm{cav}}$ = Emitting cavity drive power, **Rec. len.** = length of the recorded time trace(s), **Rec.** SP = Recording span, **Monitors** = Online monitoring of $\mathbf{R}$ = Reflected power, $\mathbf{F}$ = Forward power, $\mathbf{T}$ = Temperature of the cavities.

Date	Pub.	B_{mag} [T]	P_{cav} [W]	Rec. len [h]	Rec. SP [kHz]	Test tone	Monitors
2/12	[101]	$\times$	37	11.5	0.02	$\times$	R
06/12	[102]	0.5	7	4, 4, 4	2	$\times$	R
12/12	$\times$	$\times$	45	29	2	$\checkmark$	R
4/13	[103]	2.8	33	16, 14	2	$\checkmark$	R, F, T
6/13	[104]	2.8	43	12, 10	2, 20	$\checkmark$	R, F, T
9/13	[104]	$\times$	36	29, 29, 29	2	$\checkmark$	R, F, T

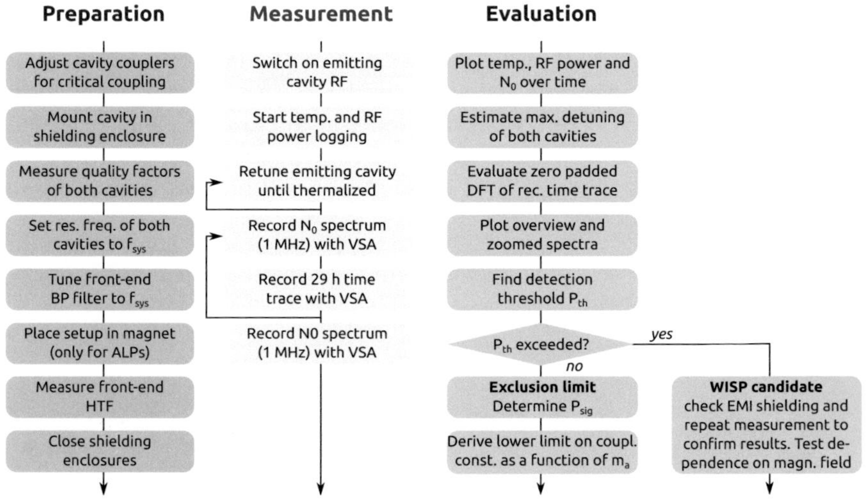

Figure 5.2.: Flow graph, showing the typical steps during the WISP measurement process.

5.1. Axion Like Particles

5.1.1. Measurement

For ALPs, the most sensitive run was carried out in July 2013 in cooperation with the Brain & Behaviour Laboratory of Geneva University. We were able to operate the setup within the bore of a 3 T superconducting magnet, which is part of an MRI scanner. Over the course of one weekend, we recorded 2 x 10 h of experimental data. A picture of the magnet and the

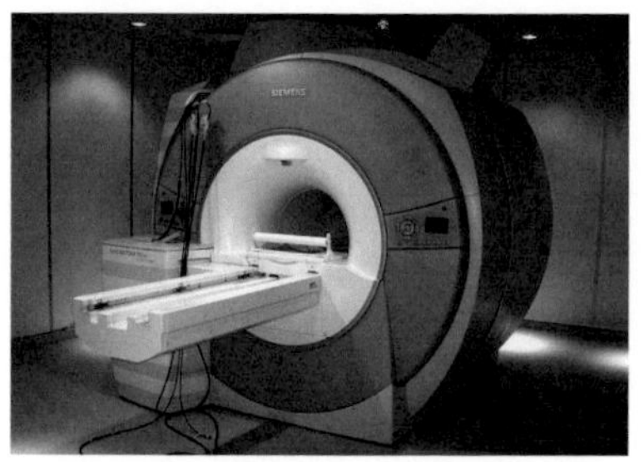

(a) 3 T solenoid magnet.

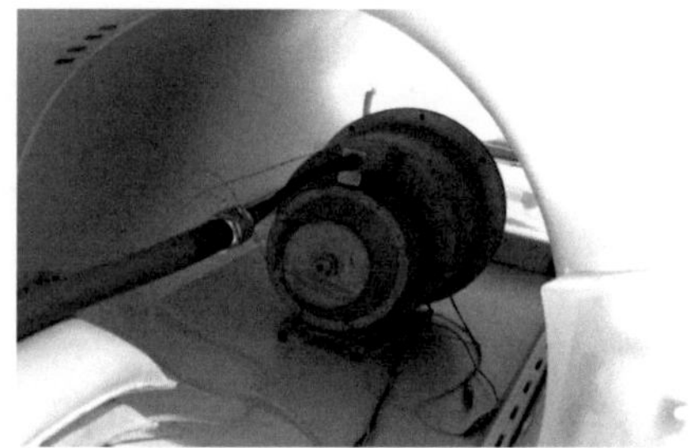

(b) Setup within the magnet.

Figure 5.3.: ALP measurement within the superconducting magnet of an MRI scanner at the Brain & Behaviour Laboratory of Geneva University.

experimental inset is shown in Fig. 5.3. The magnet provides a solenoid like field, which made it necessary to place the cavities on their sides. Care had to be taken while installing the experimental inset, as the magnet cannot be ramped down[1]. Any object with significant ferromagnetic properties will be attracted with great force towards the magnet and can quickly turn into a projectile and hence into a serious safety hazard. Several non-magnetic tools have been custom made, i.e., for tightening SMA connectors. The VSA shielding box and the RF power amplifier have been placed in a different room, behind the magnet. Feedthrough ports in the wall allowed

Table 5.2.: Parameters of the ALP run in June 2013

$f_{\mathrm{sys}} = 1.739990$ GHz	$Q_{\mathrm{det}} = 11392$	$Q_{\mathrm{em}} = 12151$	$B = 2.88$ T
$P_{\mathrm{sig}} = 9.8 \cdot 10^{-25}$ W	$P_{\mathrm{em}} = 47.9$ W	$|G|_{\mathrm{max}} = 0.94$	Mode $= \mathrm{TM}_{010}$

us to place the optical fibres and RF cables. The technical parameters of the run have been summarized in Table 5.2.

For this particular measurement run, two signal receivers have been used in parallel. The Agilent VSA recorded the signal from the cavity with 2 kHz bandwidth and the custom downmixing chain with 20 kHz bandwidth. The analog signal was split by a 3 dB hybrid, as indicated in Fig. 3.26. Its attenuation has been taken into account during the calibration with the HTF.

Recording the data with the custom downmixing chain proved to be problematic. Due to constraints in space and thermal dissipation, it was not possible to place an additional PC into the receiver shielding enclosure. The only available PC was part of the VSA, running the Windows XP operating system. It was found that the CPU load during VSA data acquisition is very low, hence it was decided to also use this PC to record data from the soundcard of the custom downmixing chain. At this time, it was not known that the Windows XP operating system has the tendency to drop audio samples, as explained in 4.6.2.

5.1.2. Cross checks

Figure 5.4(a) shows a trace of $P_{\mathrm{em}} = P_{\mathrm{fwd}} - P_{\mathrm{ref}}$, which is the absorbed RF power of the emitting cavity, taking reflection due to mismatch into account. During this run, an unexpectedly large fluctuation of the ambient temperature resulted in a thermal runaway condition after the first 12 h of data taking. The emitting cavity drifted off-resonance, reflected all

---

[1] The field of the magnet can only be ramped down by a quench in case of emergencies. During normal operation, a superconducting switch short-circuits the coil, keeping the magnet energized indefinitely as resistive losses are negligible. This mode of operation has advantages for MRI systems, as the magnetic field is extremely stable over time.

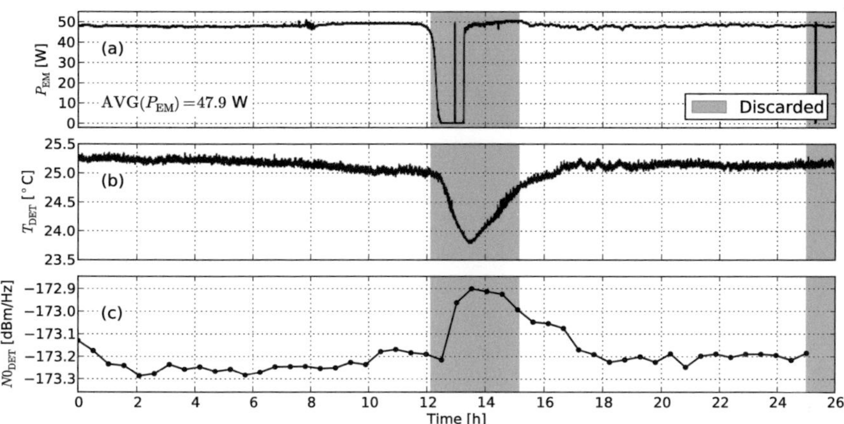

Figure 5.4.: All data recorded during the measurement run in June 2013
(a) Absorbed RF power of the emitting cavity, taking mismatch into account.
(b) Temperature of the detecting cavity shielding box. **(c)** Thermal noise
power density from the detecting cavity, measured in a bandwidth of 2 kHz
around f_{sys}.

incident RF power and cooled down to ambient temperature within a few
minutes.

After noticing this condition, it was possible to bring the cavity back to
the nominal operating temperature and resonant frequency by adapting
f_{sys} remotely. We were able to continue the experiment after a 3 h period,
during which the recorded data had to be discarded. Despite the thermal
runaway incident, this run still yields highest sensitivity towards ALPs.

The detecting cavity was only tuned at the beginning of the measurement
run. After closing its shielding enclosure, the tuning screw is not reachable
and was left untouched. A good indication of its resonant frequency at the
beginning (t=0 h) and end (t=25 h) of the measurement run is given by the
maximum of its spectral noise power density (N_0) in Fig. 5.5. The cavities
f_{res} is $\approx$ 20 kHz below f_{sys}. The noise power at f_{sys} is $\approx$ 0.3 dB below
the maximum. We would expect the same amount of degradation for a

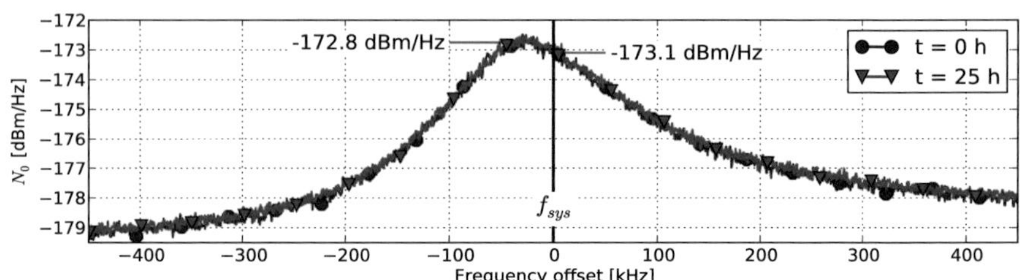

Figure 5.5.: Noise power density at the coupling port of the detecting cavity, indicating its resonant frequency in relation to the system frequency f_{sys}. The measurement has been done immediately before and after the 25 h ALPs measurement run in June 2013. No significant drift was observed.

hypothetical WISP signal. There is no visible difference between the state of the cavity at the beginning and end of the measurement run. However, the temperature measurement, shown in Fig. 5.4(b) indicates a significant change of $\Delta T = -1.25°$C at t=12 h. The cause for this fluctuation was the unexpected cool-down of the emitting cavity, which was in close vicinity to the detecting cavity. According to Fig. 3.10, the change in temperature corresponds to a relative change in resonant frequency of $\approx \Delta f = +40$ kHz, which would place the cavities resonance ≈ 20 kHz above f_{sys} and cause a worst case reduction of -0.3 dB in signal power. Note that it was not practical to measure the actual temperature of the detecting cavity, as its EMI shielding enclosure would have been compromised by the copper wires of the temperature sensor. Instead, the sensor has been placed on the outside wall of the shielding enclosure, which is in good thermal contact with the cavity. The actual fluctuations of cavity temperature are therefore less than the measured T_{DET}.

As a further crosscheck, the average noise power in a bandwidth of 2 kHz around f_{sys} has been evaluated from the recorded experimental data. The result is shown in Fig. 5.4(c). The noise power density at t=0 h is $N_0 = -173.1$ dBm/Hz, which is in good agreement with the blue trace in Fig. 5.5. At t=12 h, an excursion of +0.3 dB is visible. This indicates a shift of the resonance curve by $\approx \Delta f = +20$ kHz, centering it on f_{sys}. At t=18 h, the cavity has warmed up again and reached its original and slightly detuned state, which is in good agreement with the green trace in Fig. 5.5.

In conclusion, we can make the following statements for the ALPs run in June 2013:

- The worst case signal degradation of an hypothetical ALP signal due to detuning of the detecting cavity was ≤ 0.3 dB $= 7\%$.

- The average RF power absorbed by the emitting cavity was $P_{\mathrm{em}} = 47.9$ W

5.1.3. Spectra

During the HSP and ALP measurement runs, the center frequency of the VSA was set to $f_{\mathrm{center}} = f_{\mathrm{sys}} + 400$ Hz to avoid internal spurious signals appearing at interesting parts of the spectrum [101]. The span was set to $SP = 2$ kHz, which provides enough bandwidth to resolve spurious signals and to establish sufficient distance between the WISP and test tone signals.

The final spectrum in Fig. 5.6 has been derived by a DFT operation of the zero padded time trace, according to 4.2.

Note that the test tone signal appears as a very sharp peak in the spectrum. Therefore it is demonstrating that the frequency drifts between the RF sources and the VSA are negligible and the narrowband detection concept is working.

As a crosscheck, we can derive the expected average noise power in one bin (P_N') by

$$P_N' = k \cdot BW_{\mathrm{res}} \cdot (T_{sys} + T_{cav}), \qquad [\mathrm{W}] \tag{5.1}$$

where k is the Boltzmann constant and $BW_{\mathrm{res}} = 1/l = 28.2\ \mu\mathrm{Hz}$ is the resolution bandwidth, which is inversely proportional to the length of the recorded time trace l. The noise temperature of the receiving chain and of the detecting cavity is T_{sys} and T_{cav} respectively. Note that the former is related to the noise figure from the HTF measurement ($NF = 0.7$ dB, $F = 1.17$) by: $T_{sys} = 290 \cdot (F-1) = 50.7\ K$. The noise temperature of the

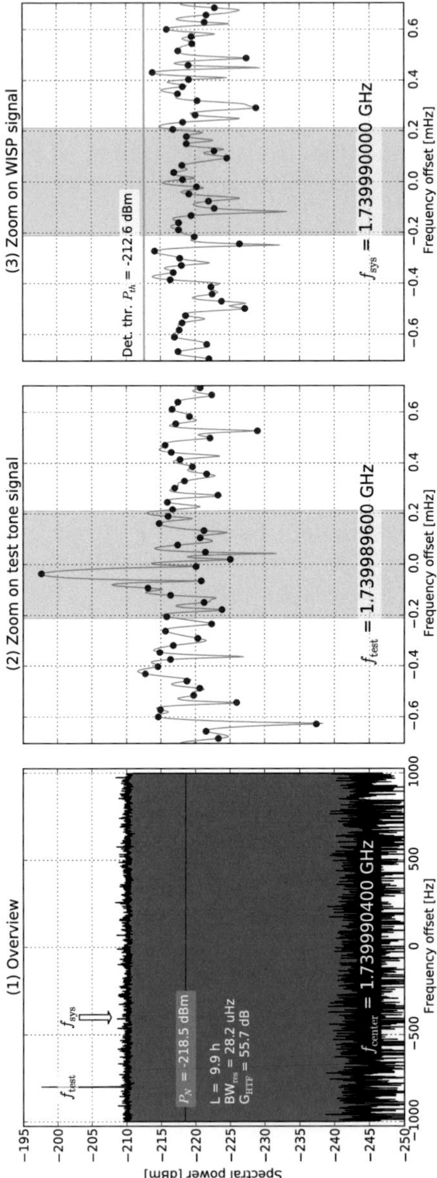

Figure 5.6.: Resulting power spectrum from the ALP run in June 2013. (1) Overview over full recorded span. (2) Zoom on f_{test}, where the test tone signal is clearly visible. (3) Zoom on f_{sys}, where no WISP signal is visible. The green shaded areas mark the frequency range where a signal would be expected. Blue dots indicate the spectral bins calculated by a direct DFT operation, the grey lines show the underlying continuous spectrum, approximated by zero padding the time domain data. All three plots share the same Y - axis.

detection cavity is equivalent to its physical temperature on its resonant frequency: $T_{cav} = 299.4$ K. The expected P'_N from these values is in good agreement with the measured P_N:

$$P'_N = 1.36 \cdot 10^{-25} \text{ W} = -218.6 \text{ dBm} \qquad P_N = -218.5 \pm 0.5 \text{ dBm}. \tag{5.2}$$

The uncertainty in the measured value originates from the measurement error of the HTF - Gain parameter, which was used to normalize the spectra to absolute power at the detecting cavities coupling port.

The detection threshold has been determined from the statistics of the background noise by the method described in 4.4. The minimum detectable power with a 95 % confidence level, taking a 0.3 dB degradation of signal power due to the detuning of the detecting cavity into account, is $P_{sig} = 9.8 \cdot 10^{-25} \text{W} = -210.1$ dBm.

5.1.4. Exclusion limits

An upper bound on the axion coupling constant to photons (g) has been derived according to Eq. 2.15 from the parameters in Table 5.2. The exclusion plot is shown in Fig. 5.7. For ALPs, the experiment was – in a small mass range – more sensitive than other purely laboratory based experiments (namely laser LSW of the first generation like ALPS-1 [46]) but less sensitive than extraterrestrial experiments like CAST [47]. This is the first time ALPs have been probed by a microwave based LSW experiment.

5.2. Hidden Sector Photons

5.2.1. Measurement

At the time of writing, the most sensitive measurement run for HSPs has been carried out in September 2013 at CERN. We recorded three continuous time traces, each of 29 h length, which have been averaged

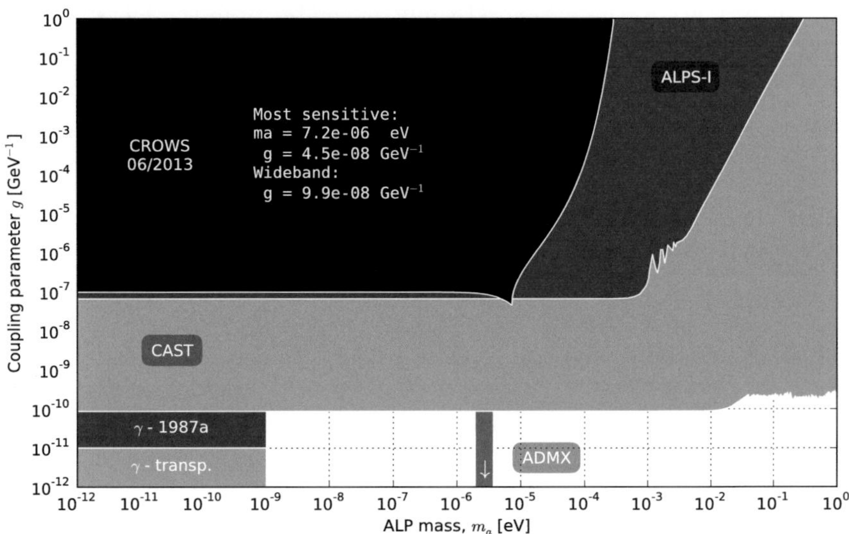

Figure 5.7.: Exclusion limits for ALPs for the measurement run in June 2013 in a 3 Tesla magnet. Confidence level: 95 %. Laser LSW = exclusion limits from ALPS-1, the most sensitive optical LSW experiment to date [46]. Note that CAST [47] is a helioscope, searching for solar ALPs and is thus not a purely terrestrial laboratory experiment. More details on the other experiments can be found in [26].

during data evaluation to achieve the lowest background noise level of all runs.

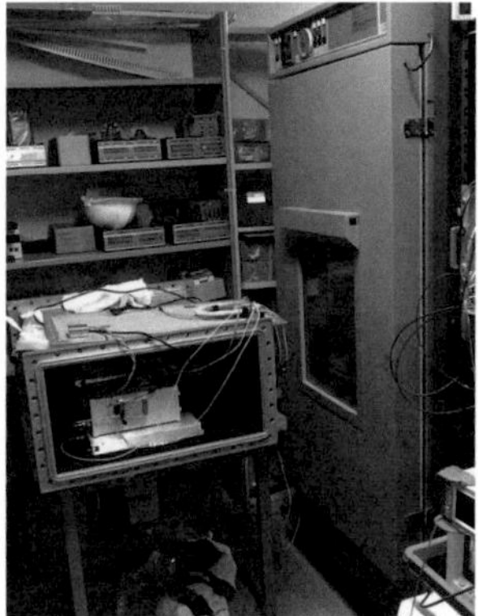

(a) Principle arrangement of the components, as they have been placed in the temp. controlled chamber.

(b) Open VSA shielding box set up next to the temp. controlled chamber.

Figure 5.8.: The HSP measurement in September 2013 has been carried out in a temperature controlled chamber, which provided a stabilized environmental temperature of 17.2 °C and allowed us to operate the setup for 100 h.

Earlier HSP measurements were carried out in a laboratory at CERN without any facilities for environmental temperature stabilization. We found that during longer recordings, the temperature change from day- to nighttime had a significant influence on the cavities resonant frequency. To eliminate these fluctuations, the final HSP run has been carried out in a temperature regulated chamber. It provides a steady air flow of heated or chilled air and was set to stabilize the environmental temperature to 17.2 ± 0.5 °C. The temperature fluctuations within the chamber during the 100 h run have been measured and are shown in Fig. 5.9(b). A picture of the setup is shown in Fig. 5.8.

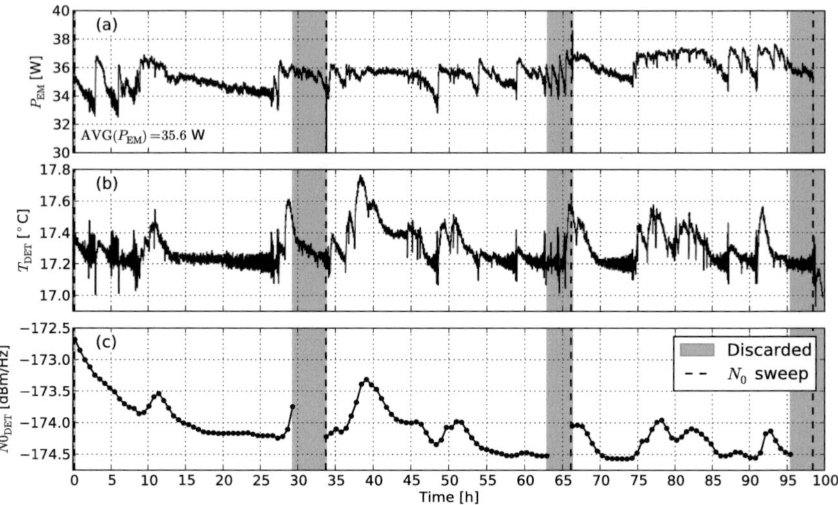

Figure 5.9.: (a): emitting cavity absorbed RF power (defined as $P_{\mathrm{EM}} = P_{\mathrm{inc}} - P_{\mathrm{refl}}$), (b): Physical temperature of the detecting cavity, (c): Noise power density from the detecting cavity within a 2 kHz bandwidth around f_{sys}. The vertical black dashed lines indicate when a dedicated N_0 sweep has been taken (shown in Fig. 5.10).

5.2.2. Cross checks

As opposed to the ALP run described in 5.1, no thermal runaway condition occurred during the latest HSP run. Figure 5.9(a) shows that the RF power amplifier and the emitting cavity were stable during the entire 100 h of continuous operation at full RF power. Around 14 W of RF power was lost due the 50 cm long coaxial cable, connecting the power amplifier outside the thermal chamber to the emitting cavity within.

While the detecting cavity was perfectly on tune in the beginning of the measurement run (blue trace in Fig. 5.10), a significant drift took place during the following 15 h. The drift can be observed in the first quarter of Fig. 5.9(c), showing the noise power density (N_0) from the detecting cavity over time and by the green trace in Fig. 5.10, showing N_0 over

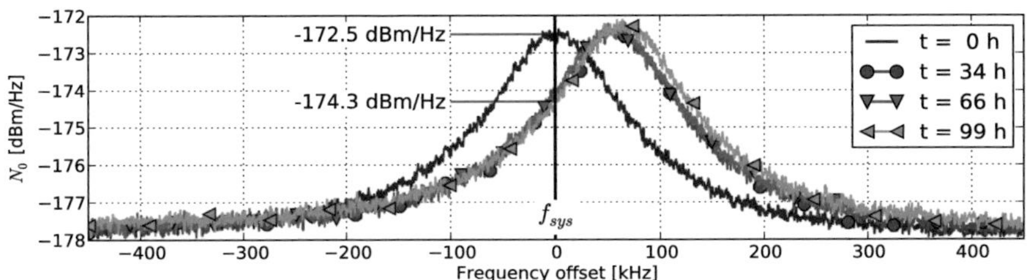

Figure 5.10.: Noise power density N_0 at the coupling port of the detecting cavity, indicating its resonant frequency in relation to f_{sys}. A new trace has been recorded at the beginning and end of each data taking interval. A significant drift of the cavity was observed, which will attenuate a potential HSP signal by ≈ 1.8 dB.

frequency at t=34 h. The drift cannot be observed in the temperature trace in Fig. 5.9(b), which has been measured on the outside of the detecting cavity shielding enclosure. These observations provide evidence that the detecting cavity did not have enough time to thermalize completely with its environment. At the beginning of data taking, it was still cooling down to the temperature of the shielding enclosure. Nonetheless, the tuning screw of the detecting cavity was not adjusted as the drift was noticed too late into the measurement run. Furthermore, the tuning screw is not easily reachable from within the thermal chamber and the EMI shielding enclosure. From the decrease in noise power at f_{sys} in Fig. 5.10 and Fig. 5.9(c) we can estimate that the drift in resonant frequency would attenuate a potential WISP signal by ≈ 1.8 dB, which has been taken into account for the exclusion result. In conclusion, we can make the following statements for the HSP run in September 2013:

- The worst case signal degradation of an hypothetical ALP signal due to detuning of the detecting cavity was ≤ 1.8 dB $= 34\%$.

- The average RF power absorbed by the emitting cavity was $P_{\text{em}} = 35.6$ W

5.2.3. Spectra

The final spectrum in Fig. 5.12 has been obtained by processing each of the three recordings by a zero padded DFT operation (according to 4.2) and subsequently averaging the three resulting power spectra. The detection threshold has been determined according to 4.4 by evaluating the background noise statistics from a part of the spectrum, where no HSP signal is expected.

Note that the test tone signal can be observed with the minimum possible spectral width. It appears to be smeared out – not because of frequency drift – but because of spectral leakage. To demonstrate this, the minimum possible signal width, governed by the rectangular window function has been overlayed by a white trace. Therefore, we can state that the frequency drift between the RF sources and the VSA are negligible and the narrowband detection concept is working properly.

The spectral resolution is almost three times higher compared to the ALPs run. The window where the HSP signal is expected has been defined with a width of 400 μHz, which is the same as for the ALPs run. Due to the higher spectral resolution, it contains $T = 45$ frequency bins. The detection threshold for $\mathrm{PR}_\alpha := 5\,\%$, according to 4.4, is $P_{th} = -218.4$ dBm. It is very close to – but still above – the highest observed signal peak. Considering the implicit uncertainity of ± 0.5 dB, originating from the HTF which is used to normalize the spectrum, we do not see a statistical significant peak above the detection threshold.

As a crosscheck, we can derive the expected average noise power in one bin (P_N') in a similar fashion to 5.1.3. The relevant parameters are: $\mathrm{BW}_{res} = 9.5\ \mu$Hz, $T_{sys} = 50.7\ K$, $T_{cav} = 291.5$ K. The expected P_N' from these values is in good agreement with the measured P_N:

$$P_N' = 4.5 \cdot 10^{-26}\ \mathrm{W} = -223.5\ \mathrm{dBm} \qquad P_N = -224.1 \pm 0.5\ \mathrm{dBm}. \tag{5.3}$$

The slight discrepancy can be explained by the detuning of the detection cavity, reducing the measured noise power and attenuating a hypothetical WISP signal. The worst case sensitivity degradation of 1.8 dB due to this effect has been taken into account in the exclusion result.

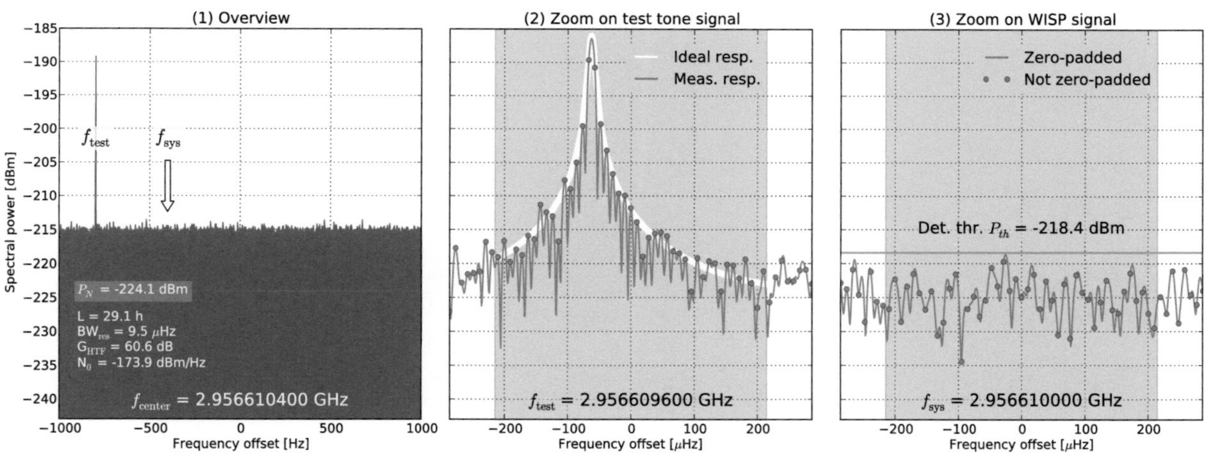

Figure 5.11.: Resulting power spectrum from the HSP run in September 2013. (1) Overview over full recorded span. (2) Zoom on f_{test}, where the test tone signal is clearly visible. Its roll-off is dictated by the rectangular window-function of the DFT, as demonstrated by the "Ideal response". (3) Zoom on f_{sys}, where no HSP signal is visible. The green shaded areas mark the frequency range where a signal would be expected. Grey dots indicate the spectral bins calculated by a direct DFT operation, the grey lines show the underlying continuous spectrum, approximated by zero padding the time domain data. All three plots share the same Y - axis.

Table 5.3.: Parameters of the HSP run in September 2013

$f_{\mathrm{sys}} = 2.956610 \ \mathrm{GHz}$	$Q_{\mathrm{det}} = 22739$	$Q_{\mathrm{em}} = 23210$	$P_{\mathrm{em}} = 35.6 \ \mathrm{W}$
$P_{\mathrm{sig}} = 3.72 \cdot 10^{-25} \ \mathrm{W}$	$\lvert G \rvert_{\mathrm{max}} = 0.51$	$\mathrm{Mode} = \mathrm{TE}_{011}$	

5.2.4. Exclusion limits

The technical parameters of the run have been summarized in Table 5.3. As no HSP candidates were observed, an upper bound for the kinetic mixing parameter of HSPs to photons (χ) has been derived from the technical parameters with Eq. 2.16. The experiment was sufficiently sensitive to improve over previous exclusion limits in its mass range, as shown by the exclusion plot in Fig. 5.12. These results have been accepted and published in the journal "Physical Review D" [104]. Therefore, at the time of writing (October 2013) the CROWS experiment is the worlds most sensitive laboratory test for HSPs in the mass range from $4 \ \mu\mathrm{eV} \leq m_{\gamma'} \leq 20 \ \mu\mathrm{eV}$.

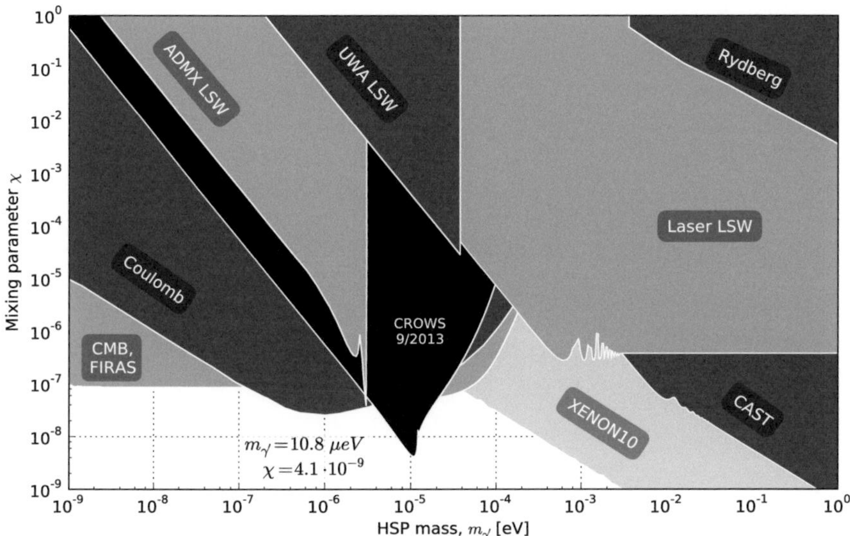

Figure 5.12.: CROWS: exclusion limits for HSPs for the measurement run in September 2013. Confidence level: 95 %. UWA LSW [63] and ADMX LSW [64] are similar microwave based LSW experiments. More details on the other experiments can be found in [26].

6. Conclusion

The QCD axion has been the most accepted solution to the strong CP problem for over 30 years now. It originated from a subtle and elegant extension to the standard model, solving a longstanding problem in theoretical particle physics. There are several motivations to search for the axion and its relatives: Axion Like Particles, Hidden Sector Photons and other WISPs. Both, ALPs and HSPs would be excellent candidates for cold dark matter, if their mass would be within a certain range. They could also solve numerous astrophysical puzzles, such as the anomalous transparency of the universe for high energy gamma rays, and arise naturally in String theory and other extensions to the SM. Nonetheless, until the present day, these WISPs are only hypothetical. Due to their weakly interacting nature, these particles are effectively hidden from high energy collider experiments. The most efficient way to probe their existence is to exploit the weak coupling to photons by the Primakoff effect (ALPs) or by kinetic mixing (HSPs). Several dedicated experiments have been proposed to convert the hidden WISP to a photon, which subsequently can be detected. These are: Haloscopes, searching for WISPs made up of dark matter; Helioscopes, searching for WISPs from the sun and Light Shining through the Wall experiments (LSW), searching for WISPs produced in the laboratory. The latter one is the only entirely self sufficient experiment. It does not depend on any external assumptions and hence can provide the most certain exclusion limits. The classical LSW experiment was based on a laser as a source of photons, shining through a magnet, exploiting the fact that some of the photons might convert to WISPs. The laser-beam is blocked by a wall, which the WISPs can penetrate to subsequently reconvert to photons in a well isolated detection volume. If one could observe a faint light appearing to be – shining through the wall – this would be an indirect indication for the existence of WISPs. The LSW principle has been adapted from the optical to the microwave regime. The main motivation is the availability of low loss microwave cavity resonators, which are much cheaper, easier to handle and more rugged than optical resonators of similar quality factor. Moreover, the lower quantum energy of microwave photons make the experiment more energy efficient. The lower frequency also makes it possible to apply coherent signal detection methods, which are not easily realized in the optical domain.

In the framework of this PhD, a microwave based LSW experiment has been developed and commissioned at CERN. Several measurement campaigns to search for HSPs and ALPs have been carried out. The latter required the experimental inset to be placed in a strong static magnetic field, which was made possible by a cooperation with the Brain & Behaviour Laboratory of Geneva University, providing us access to a 3 T superconducting magnet, being part of an MRI scanner. This was the first time, ALPs have been probed by a microwave based LSW experiment. While no ALPs were detected, similar sensitivity to "ALPS-I" was achieved, which was the most sensitive laser based LSW experiment at the time of writing. Lower bounds were set on the coupling constant $g = 4.5 \cdot 10^{-8}$ GeV^{-1} for axion like particles with a mass of $m_a = 7.2$ μeV. Several measurement runs for HSPs have been carried out in a laboratory at CERN, as no external magnetic field is necessary. The most sensitive run achieved an improvement in sensitivity by almost one order of magnitude compared to the previously most sensitive experiment in the relevant HSP mass range, which allowed us to probe a significant "unexplored" area in the exclusion plot. Lower bounds were set for the coupling constant $\chi = 4.1 \cdot 10^{-9}$ at a mass of $m_{\gamma'} = 10.8$ μeV. The experimental results for HSPs and ALPs have been published in the peer reviewed journal "Physical Review D" [104].

To make these results possible, several engineering challenges had to be solved. A cavity geometry and resonating mode had to be found, which provides a good coupling to the hidden particles, a large quality factor and which can be excited in a well defined way while avoiding crossovers with other modes. To couple the fields in the cavity to a 50 Ω system, inductive coupling antennas were designed and optimized, to accomplish less than -30 dB of return loss. The "emitting" cavity has been driven by a 50 W RF power amplifier on its resonant frequency of 1.7 GHz (ALPs) or 2.9 GHz (HSPs). Inside the cavity an electromagnetic field builds up, which is equivalent to driving a non-resonant structure by ≈ 1 MW of RF power. The detecting cavity is placed immediately next to it and tuned to the same resonant frequency. It is connected to a low noise amplifier and a sensitive receiver, which allows us to detect RF powers in the order of 10^{-24} W. One of the most critical design aspects was to prevent crosstalk between the two cavities through electromagnetic leakage. Crosstalk would lead to a signal, which cannot be distinguished from a possible WISP candidate and had to be attenuated below the detection threshold. For this reason a system of electromagnetic shielding enclosures were designed, providing about 300 dB of attenuation between the electromagnetic fields in the two cavities. The shielding had to be split into a magnetically compatible enclosure for the detecting cavity and a larger receiver enclosure, placed outside the magnet. As coaxial cables cannot provide more than ≈ 120 dB of shielding, optical fibres were used to transmit the RF signals between the two shielding enclosures. This configuration proved to be modular, easy to transport and to set up, allowing us to move the whole apparatus to laboratories where subsequently stronger superconducting magnets are available, iteratively improving the detection sensitivity for ALPs.

A necessary condition for the experiment is the agreement of the resonant frequency of both cavities with the driving signal from the power amplifier. Several methods were found and implemented to monitor the instantaneous resonant frequencies. The emitting cavity was monitored by its reflected RF power and the detecting cavity by its thermal noise density. Furthermore, the temperatures of both were recorded. This allows to monitor the detecting cavity in an entirely passive way, avoiding any kind of signals, which could potentially interfere with the WISP detection process. The measurement runs have been carried out at room temperature, so the thermal noise background is the limiting factor for signal detection. After a measurement run, the recorded time trace was evaluated for the presence of a WISP signal by determining its power spectrum. This corresponds to narrowband filtering with a resolution bandwidth down to 10 μHz. As the WISP signal is sinusoidal and its frequency is known, it can pass this narrow filter, while background noise is attenuated. By evaluating a 29 h time trace, we are capable of detecting the presence of a signal with a power of less than 10^{-24} W, which corresponds to a flux of about 0.5 photons per second at 3 GHz. The narrowband detection concept demands a relative frequency stability in the order of a few μHz for each microwave oscillator involved in the transmission and receiving chain. We were able to fulfill these requirements by the consequent use of phase locked oscillators in every stage of the receiving chain. Hence all signals are synchronized to a common reference clock.

The experimental inset has been placed in a 3 T magnetic field for the ALP runs and the electronic components of the front-end, like the low noise amplifier and the analog optical link were found to be sensitive to such an environment. This required in situ characterization of their gain and noise figure parameters. Furthermore, the initial optical transmitter was permanently damaged after the first tests in the magnet. As no commercial supplier could provide us a tested and specified solution, some experimentation was needed to find a configuration which could operate reliably in a strong magnetic field. In the case of the optical link, its circuit had to be modified to avoid magnetic components like ferrite cored inductors. Furthermore, all metallic parts of the experimental inset had to be made of non-ferromagnetic materials like copper, brass or aluminium.

A custom made downmixing chain has been designed and built, allowing us to record the detecting cavity signal with up to 20 kHz bandwidth for an unlimited amount of time. Although the receiver has not been used for the most sensitive exclusion limits shown in this work, it still is an interesting design study for a potential wideband receiver for future WISP experiments. Its biggest advantage is its ability to track frequency swept RF sources, which would allow to exploit accelerating cavities as parasitic hidden photon emitters. The frequency tracking would enable us to process the data by the narrowband receiver concept.

One step of the signal evaluation involves the calculation of a very large DFT over $> 10^{10}$ recorded samples. A software framework has been implemented, which is able to achieve this calculation on ordinary PCs. It has been thoroughly tested for bugs and numerical correctness and proved to be very useful for processing the long time domain recordings.

7. Outlook

Several opportunities for measurement runs in strong magnets were found, which could not all be carried out due to time constraints. These open points shall be discussed in this section.

There are several options to improve the sensitivity of the experiment. For ALPs, one might use a stronger magnet or lower frequency cavities, which can improve sensitivity considerably. The experimental inset of CROWS was designed to fit into a 30 cm bore of a solenoid, which makes it compatible with a large number of MRI magnets. As 7 T systems become increasingly common, it might be an option to collaborate with an institute, where such a device is available. It has been demonstrated that due to the transportable nature of the experiment, such a measurement at an external partner is feasible. In principle, no changes would be needed for the current setup. Nonetheless, the front-end electronics should be tested and characterized in the 7 T field. Sensitivity[1] of an ALP experiment scales proportionally with the magnetic field and hence could be improved by a factor of ≈ 3.

Another option is to use larger cavities at lower frequencies. Due to the larger detection volume, sensitivity would be improved. One might think of a large scale LSW experiment with recuperated cavities from an particle accelerator. For example, the discontinued 200 MHz standing wave cavities of the Super Proton Synchrotron [105] (SPS), which are basically available for free at CERN and DESY. These devices have a diameter of ≈ 2 m and would fit, with a slight modification, in the CERN M1 magnet [106], providing a 3 T solenoid field. The magnet has recently been renovated and is now available for general purpose use at CERN. The receiving chain from the CROWS experiment could be readily used for signal detection. The cavities are already attached to a powerful (60 kW)

---

[1] As given by the lower limit on the coupling constant g in Eq. 2.15.

tube based RF amplifier – but a new power converter would be necessary, which is estimated to cost ≈ 500 k CHF [107]. As sensitivity to ALPs scales with $1/f_{\mathrm{sys}}$ and with the fourth root of the excitation RF power, this configuration could improve sensitivity by a factor of ≈ 50, compared to the CROWS results presented in this work.

For hidden photon search, sensitivity is largely dependent on the quality factor of the cavities. The strongest improvement could be achieved by making use of superconducting technology on the emitting and detecting side. This is possible, as no external magnetic field is necessary for HSP search. Nonetheless, significant engineering challenges need to be overcome. The biggest problem is to keep both cavities on the same resonant frequency, which becomes increasingly difficult for higher quality factors. This configuration has been discussed in 3.2.1. As sensitivity^2 towards HSPs scales with the square-root of the cavity quality factor, we could achieve an improvement of three orders of magnitude, if the technical challenges can be mastered and quality factors in the order of $Q_L = 10^{10}$ can be accomplished.

Another interesting idea is to parasitically exploit the cavities of particle accelerators as HSP emitters. One example are the RF cavities of the SPS, which are driven by up to 500 kW of RF input power and hence build up a strong electromagnetic field. If HSPs exist, these cavities would radiate them during operation. A well shielded detecting cavity with an identical resonant frequency would have to be placed next to the accelerating cavity. The custom made downmixing chain, as described in 4.6 would be an ideal solution for data recording, as it allows us to track the frequency sweep of the accelerator.

It shall be noted here that the measurement setup could easily be modified into a Haloscope, to search for dark matter axions in an ADMX or YMCE like experiment. For the modification, the emitting cavity is removed, the detecting cavity is placed into a magnet and the recorded signal from the receiver is evaluated for unexpected sinusoidal signals. The shielding enclosure around the detecting cavity would effectively reject EM interference – however, for a Haloscope configuration, the thermal background noise cannot be reduced by narrowband filtering, as the frequency of the WISP signal is unknown. Furthermore, the cavity needs to be tunable over a

---

^{2}As given by the lower limit on the coupling constant χ in Eq. 2.16.

wide frequency range – as in contrast to LSW, Haloscope experiments are inherently narrowband and only sensitive in a WISP mass range corresponding to the cavity bandwidth. To compensate for this disadvantage and to improve the scanning speeds, several modes in the cavity could be used in parallel for axion detection.

The CROWS experiment was not only a proof of concept for the microwave LSW principle, which has been achieved with a relatively small amount of time and manpower (compared to other WISP experiments) – but it also produced competitive results and even improved over existing ones for HSPs, demonstrating the power of its approach. Appropriate solutions to all engineering problems, like cavity design, EM shielding, operational monitoring, signal processing and design for magnetic compatibility have been found. It can be expected that this pioneering work will be of value for the design and implementation of future microwave based LSW experiments at CERN and other laboratories.

Appendix

A. Cavity Theory

Lossless microwave cavities support an infinite number of resonating modes towards higher frequencies. The response of each mode around its resonant frequency can be modelled by an electrical circuit consisting of parallel (or series) R, L and C components, as shown in Fig. A.1a.

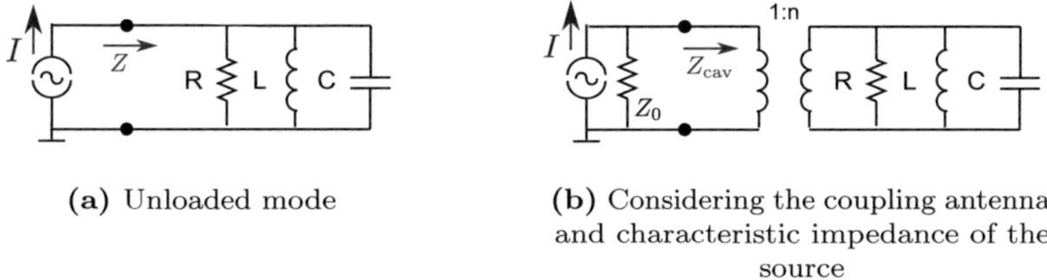

(a) Unloaded mode

(b) Considering the coupling antenna and characteristic impedance of the source

Figure A.1.: Equivalent circuit model of a resonating mode in a cavity

If we assume a RLC parallel circuit, the input impedance is defined as

$$\frac{1}{Z} = \frac{1}{R} - \frac{j}{\omega L} + j\omega C. \tag{7.1}$$

We assume the circuit is excited by a sinusoidal current of amplitude[3] I.

---

[3] The amplitude of a sinusoidal signal is not to be confused with the RMS value, which is $I_{RMS} = \frac{I}{\sqrt{2}}$.

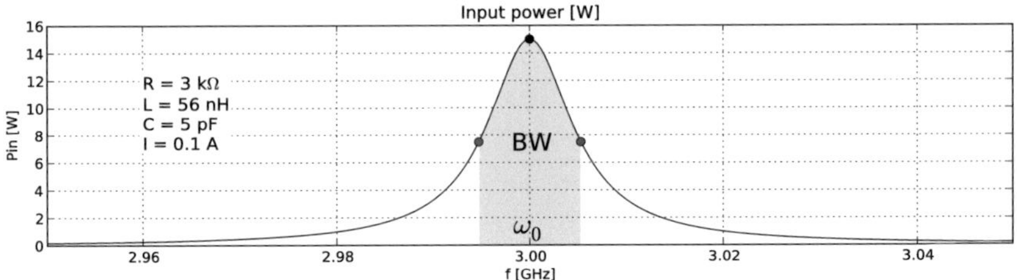

Figure A.2.: Example of the real input power of a parallel RLC circuit driven by a sinusoidal current.

The input power, absorbed by resistive losses in the cavity walls, which corresponds to R in the RLC circuit, is given by

$$P_{in} = \mathrm{Real}\left\{\frac{I^2 Z}{2}\right\}. \tag{7.2}$$

It is plotted for an example resonator in Fig. A.2. Note how P_{in} becomes maximum at the resonant frequency ($\omega = \omega_0$) and the imaginary parts cancel out:

$$P_{in} = \frac{I^2 R}{2} \quad \text{at} \quad \omega_0 = \frac{1}{\sqrt{LC}} \tag{7.3}$$

The quality factor Q_0 of this resonator is defined as average stored power divided by average power loss per cycle. Alternatively, it is defined as reactive power (power through C or L) divided by effective power (power through R) on the cavity's resonant frequency:

$$Q_0 = \frac{\frac{R^2 I^2}{2\omega_0 L}}{\frac{I^2 R}{2}} = \frac{R}{\omega_0 L} = R\omega_0 C \tag{7.4}$$

For small deviations from the resonant frequency ($\frac{\Delta\omega}{\omega_0}$) and with the definition of Q_0 and ω_0, we can construct a simplified formula for the cavity's input impedance:

$$Z \approx \frac{R}{1 + j2Q_0 \frac{\Delta\omega}{\omega_0}} \qquad (7.5)$$

The 3 dB bandwidth BW of the mode is defined as the distance in frequency between the two points (marked red in Fig. A.2), where the cavity only absorbs half of the input power. It is related to the quality factor (Q_0) by

$$BW = \frac{\omega_0}{2\pi Q_0} \qquad [\text{Hz}] \qquad (7.6)$$

Note that Q_0 is the unloaded quality factor. It does not take any external losses into account, which inevitably arise from coupling the cavity mode to an external system.

Let us take the emitting cavity of the experiment as an example. We want to couple the mode in the cavity to the output of the RF power amplifier. We know that the characteristic impedance of the amplifiers output port is $Z_0 = 50\ \Omega$. An adjustable coupling antenna within the cavity is used to match this to the electromagnetic field in the cavity. A simple way to allow for this in the equivalent circuit is an ideal transformer with adjustable turns ratio n, as shown in Fig. A.1b. The amount of power absorbed by the cavity on its resonant frequency is now dependent on n (which relates to the cross section of the antenna) and needs to be maximized.

The effective cavity impedance as seen through the transformer (or coupling antenna) is given by:

$$Z_{\text{cav}} = \frac{Z}{n^2} \approx \frac{R}{n^2 \left(1 + j2Q_L \frac{\Delta\omega}{\omega_0}\right)} \approx \frac{\beta Z_0}{\left(1 + j2Q_L \frac{\Delta\omega}{\omega_0}\right)}. \qquad (7.7)$$

Maximum power is absorbed in Z_{cav} for the impedance matching condition $Z_{cav} = Z_0^* = 50\Omega$. That means, to get maximum power transfer on the resonant frequency, we need to set n to the following:

$$\frac{R}{n^2} = Z_0 \quad \rightarrow \quad n = \sqrt{\frac{R}{Z_0}}. \tag{7.8}$$

For cavity resonators, instead of the turns ratio of a virtual transformer, the coupling factor β is used. It is related to the **parallel** RLC equivalent circuit by

$$\beta = \frac{R}{n^2 Z_0}. \tag{7.9}$$

If $\beta < 1$, the cavity is undercoupled, if $\beta > 1$, the cavity is overcoupled. Maximum power is absorbed for $\beta = 1$, then the cavity is critically coupled.

Note that coupling the cavity to the amplifier has reduced its quality factor. This is due to losses in the internal resistance of the amplifier, which add to the losses in R. The resonator is loaded by the external circuit. To get the loaded quality factor (Q_L), we can move Z_0 to the right side of the transformer and get $n^2 Z_0$. Considering it is parallel to R now and with Eq. 7.9 we get:

$$Q_L = \frac{1}{1 + \beta} Q_0. \tag{7.10}$$

Equation 7.10 shows that for very weak coupling, resonator loading is small and there is no significant difference between Q_L and Q_0. Often the Q_0 value of a resonator is of interest – however it cannot be measured directly. Nonetheless, Q_L can be measured with a small coupling antenna ($\beta \approx 0$), which allows us to estimate Q_0.

As we will connect the cavity coupler to a coaxial transmission line with a characteristic impedance of $Z_0 = 50\ \Omega$, we can derive its reflection

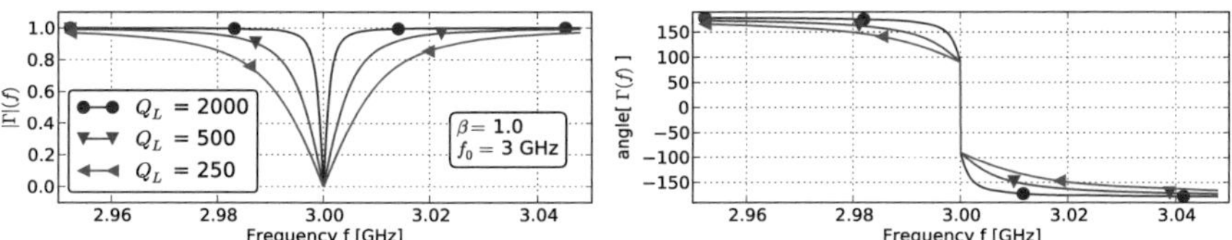

Figure A.3.: Illustration of Eq. 7.11: Magnitude and phase of the input reflection coefficient as a function of frequency and loaded quality factor for a critically coupled cavity.

coefficient Γ in that system. It is related to the impedance of the resonator Z_{cav} by (with Eq. 7.7):

$$\Gamma = \frac{Z_{cav} - Z_0}{Z_{cav} + Z_0} \approx \frac{2\beta}{\beta + 2jQ_L\frac{\Delta\omega}{\omega_0} + 1} - 1. \qquad (7.11)$$

The equation has been plotted in Fig. A.3 for a typical cavity configuration. It can be seen that for a critically coupled cavity ($\beta = 1$), the reflection at its input port becomes zero. Depending on the quality factor, the response curve of the cavity is more or less steep.

Note that we can relate these quantities to real life measurements by taking into account, that each cavity is only equipped with a single coupling port. Therefore the reflection coefficient is identical to the scattering parameters ($\Gamma = S_{11}$), which can be measured with a Vector Network analyzer (VNA) in a straight forward manner.

Moreover, the reflection coefficient can be related to the forward (P_{fwd}) and reflected (P_{refl}) RF power at the coupling port by:

$$|\Gamma|^2 = \frac{P_{\text{refl}}}{P_{\text{fwd}}}. \qquad (7.12)$$

This is a useful relationship, which has been exploited for estimating the emitting cavities resonant frequency by measuring the reflected RF power in 3.3.1.

B. Noise in the receiver chain

The charge carriers (electrons or holes) in every resistive element, therefore in every resistor, conductor or semiconductor are governed by random motion if they are at non-zero temperature. Taking a resistor as an example: a small voltage can be measured across its terminals due to this "thermal noise". The RMS value is given by:

$$V_n = \sqrt{\frac{4\,h\,f\,BW\,R}{e^{hf/(kT_n)} - 1}}, \qquad (7.13)$$

where $h = 6.626 \cdot 10^{-34}\,Js$ is Planck's constant, f is the frequency, BW the measurement bandwidth, R the resistance, $k = 1.38 \cdot 10^{-23}\,J/K$ is Boltzmann's constant and T_n is the physical temperature of the resistor. Equation 7.13 is generally known as Planck's blackbody radiation law. The predicted noise voltage is finite, even for very high frequencies, avoiding an "ultraviolet catastrophe".

To model thermal noise in a resistor, we can connect a voltage source of value V_n in series or a current source of value I_n in parallel with it. In case of a passive circuit with several resistors, all noise sources will be uncorrelated and therefore the squared averages of the noise voltages (or currents) can be added (corresponding to adding up the average noise power).

In the microwave regime we can safely assume $hf << kT_n$, drop the frequency dependence and simplify Eq. 7.13 to:

$$V_n = \sqrt{4\,k\,T_n\,BW\,R}, \qquad P_n = \frac{V_n^2}{4R} = k\,T_n\,BW, \qquad (7.14)$$

where P_n is the available (or maximum extractable) noise power from the resistor. It is common to normalize these results to 1 Hz bandwidth to get a noise power **density** N_0 with the unit of [W/Hz] or [dBm/Hz]. Often a noise temperature is quoted, which is simply related to N_0 by the Boltzmann constant:

$$N_0 = k\,T_N = 4 \cdot 10^{-21} \text{ W/Hz} \approx -174 \text{ dBm/Hz.} \qquad (7.15)$$
$$\text{(at } T_N = 290 \text{ K} = 16 \text{ °C)}$$

B.1. Noise of the cavity

We use the spectral noise power from the detecting cavity to estimate its
resonant frequency during an experimental run. We have seen in A that
the cavity can be modeled by an RLC parallel circuit. Only the R part
contributes to the noise at the cavities coupling port, however L and C
make it frequency dependent. To consider its thermal noise, a current
source needs to be put in parallel to R with a value of I_N. The thermal
noise of the complex cavity impedance Z_{cav} is then:

$$I_n = \sqrt{\frac{4\,k\,T_n\,BW}{\mathrm{Real}\,(Z_{cav})}}. \tag{7.16}$$

Note how only the real part of the impedance determines the noise current.
In general, only a component which shows resistive losses does also emit
thermal noise.

Considering the reflection factor, we can derive the noise power density at
the cavities coupling port to be:

$$N_0 = k\,T_N = \left(1 - |\Gamma|^2\right)\,k\,T_{cav}, \tag{7.17}$$

where T_{cav} is the physical temperature of the cavity. It is interesting to note
how the noise temperature T_N is equivalent to the physical temperature of
the critically coupled cavity at its resonant frequency (where its reflection
factor is very close to zero). Off-resonance, noise power decreases, effectively
following the cavities response curve.

B.2. Amplifiers

Each component of the RF front-end (described in 3.5) adds noise to
the signal from the cavity. We are motivated to minimize that added
noise as it decreases the sensitivity of the experiment. One example is a
microwave amplifier, for which the Noise Figure (NF) is often quoted in
the data sheet. Its signal and noise power at the input port is amplified,
proportionally to the gain factor G. However, due to the intrinsic noise of

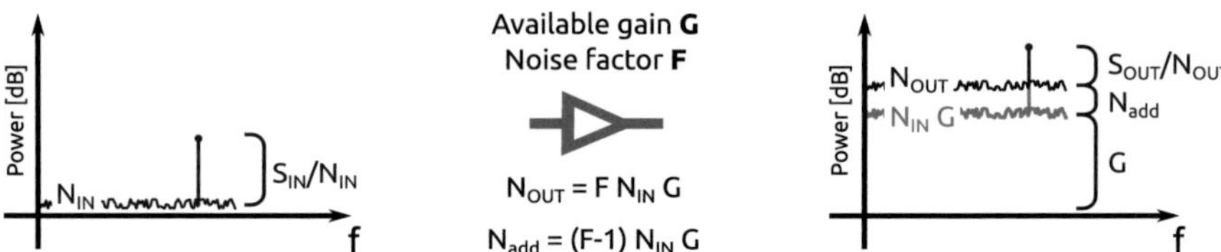

Figure B.4.: The noise factor F of an amplifier is defined as the degradation of the signal to noise ratio at its output due to its intrinsic added noise N_{add}.

the amplifier, there is an additional noise component N_{add}, degrading the signal to noise ratio at its output port.

$$F = \frac{S_{in}/N_{in}}{S_{out}/N_{out}} = \frac{1}{G}\frac{N_{in}G + N_{add}}{N_{in}} = 1 + \frac{N_{add}}{GN_{in}}, \qquad (7.18)$$

$$NF = 10\log\left(F\right).$$

Note that passive microwave components (e.g., attenuators) have a noise figure which is identical to their attenuation value.

A popular way to model the noise from an amplifier is its equivalent noise temperature T_n. The amplifier is considered as noiseless, but a resistor with a certain noise temperature is connected to its input. It is chosen to produce the same noise power at the output of the noiseless amplifier. Noise factor (F) and T_n of the amplifier are interchangeable properties and related by:

$$T_n = (F - 1)\,290\ K. \qquad (7.19)$$

If there are several amplifiers or other microwave components in a chain, the cascaded Noise factor (F_{sys}) or noise temperature (T_{sys}) of the whole chain can be calculated by Friis formula:

$$F_{sys} = F_1 + \frac{F_2 - 1}{G_1} + \frac{F_3 - 1}{G_1 G_2} + \dots \tag{7.20}$$

$$T_{sys} = T_{n1} + \frac{T_{n2}}{G_1} + \frac{T_{n3}}{G_1 G_2} + \dots \tag{7.21}$$

It is interesting to note how the noise factor of subsequent components is divided by the gain of the first amplifier. In fact, if the first amplifier provides sufficient gain, the noise factor of all other components in the receiving chain can be neglected.

The Noise figure, e.g., of an amplifier or of a complete receiving chain can be measured by the Y-factor method. A calibrated noise source is connected to the amplifiers input. The noise temperature needs to be switchable between two accurately known values, T_H and T_C. This can be realized by two matched load resistors at two different physical temperatures (e.g., one cooled by liquid nitrogen, one at room temperature). The excess noise ratio (ENR) of these two sources needs to be determined after Eq. 7.22. Alternatively, a specialized noise diode can be used, which often has the calibrated ENR ratio printed on its case. The Y-factor needs to be determined from the output powers (P_{hot} and P_{cold}) of the amplifier, with the hot and cold noise source connected:

$$Y = \frac{P_{Nhot}}{P_{Ncold}} \qquad ENR = \frac{T_H - T_C}{290K}. \tag{7.22}$$

Then the noise factor of the amplifier can be determined from these values by:

$$F = \frac{ENR}{Y - 1}. \tag{7.23}$$

This can be best explained by Fig. B.5. There is a linear relation between the temperature of the noise source on the amplifiers input and the noise power at its output. With the measurement we establish two points on the line, which completely determines it. We are interested in N_{add}, which

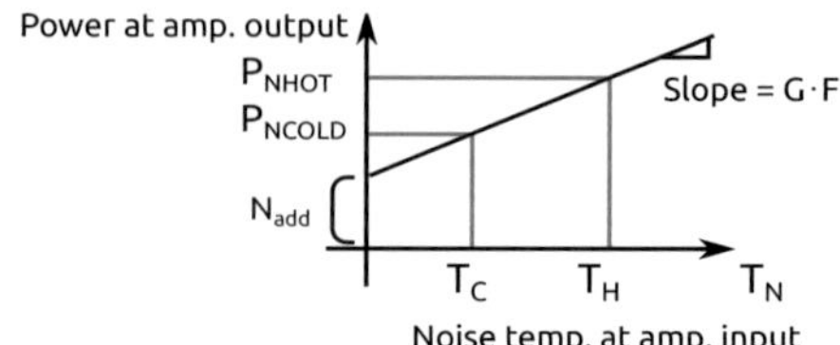

Figure B.5.: By connecting a load with two well defined noise temperatures to the amplifiers input, its noise factor (F), gain (G) and added noise (N_{add}) can be extrapolated.

is the amplifiers output noise, when there is no input noise. This cannot be measured directly, but extrapolated from the linear equation, set up by the two measurement points.

Acknowledgements

This work is the result of a collaboration between the CERN Beams-department in Geneva and the Institute for Pulsed Power and Microwave Technology at the Karlsruhe Institute of Technology (KIT). It has been supported by the Wolfgang-Gentner-Programme of the Bundesministerium für Bildung und Forschung (BMBF).

The idea of realizing a microwave based LSW experiment at CERN came from Fritz Caspers in the first place. I would like to thank him, not only for persuading the idea and making this work a reality – but also for his thoughtful supervision during the last three years. He proved to be always at hand if there was a problem, while leaving me a lot of freedom to creatively find solutions in my own way, allowing me to develop in my profession and as a person. I'm also thankful to my second CERN supervisor Marek Gasior for his constant support, for contributing many important ideas and for sharing his extensive experience in analog electronics. I would like to thank Manfred Thumm for his unconditional trust, his enthusiasm and for agreeing to supervise me from the academic side without the slightest hesitation.

During my work on the experiment, I had the chance to visit Yale university in New Haven. For the invitation to the YMCE experiment, the fruitful collaboration and exchange I would like to thank Keith Oliver Baker and the whole YMCE crew.

It required a lot of luck to find a suitable magnet which allowed to perform the most sensitive ALP measurement runs. I want to express my special thanks to the Brain & Behaviour Laboratory of Geneva University, for making their MRI magnet available to us on weekends; to Sebastian Rieger, for helping and supporting us at the MRI lab in his spare time and to Clare Burrage, for bringing the right people together at the right time.

I would like to thank Jörg Jäckel, Pascal Blanc, Felix Bergsma and Konstantin Zioutas for theoretical and practical help and a large number of inspiring and entertaining discussions. I am also very grateful for the practical assistance from the BE-RF mechanical workshop at CERN.

Thanks to the DESY crew, especially to A. Ringwald, A. Lindner, M. Schwarz and B. Döbrich, for all the organized workshops and seminars and for inviting me to actively participate in them.

I would like to express my gratitude to Ana Malagon and Manfred Wendt, for reading the manuscript and their throughout constructive criticism.

Many thanks to the colleagues from my section and the 864 coffee table, whose cheerfulness contributed to the familiar and ease working atmosphere. It was great to share the highs and lows of working on a PhD with Silke Federmann and Thomas Kaltenbacher, thanks for suffering and laughing with me.

The last three years have been a truly life-changing experience for me and it would not have been half as enjoyable without the company of my close friends. Special thanks go to Luca and Jose for rocking the office together and to Iikka, Janne and Pietro for those amazing "practical" language courses during syömään. Some of the best (and most inspiring) parts of my spare time were spent outside, exploring the nearby Alps and Jura mountains. Thanks to Moritz, Cenk, Andras, Petri and Jakub for the relaxing hiking and biking trips, the skiing outings, the marvelous paragliding jump, the rafting adventures through ice-cold water and the days of climbing on some of the most stunning rock walls in the area. Thanks to my flatmates Krish, Tom and Sebastian for memorable house parties and an excellent demonstration of their respective local cuisine: Tandori, Haggis, Buckfast and Currywurst. Thanks to Valentin, Valentina, Vely and Georgios for the wonderful summertime BBQs and coming round ever so often. Thanks to all of you and whom I might have forgotten, for making my stay in Geneva a truly unforgettable experience.

Last but not least I want to send a big "thank you" and a warm hug to my parents, Eugen and Christine, who were utterly supporting me and always will, no matter in which corner of the world I will be ending up.

Bibliography

[1] G. Aad et al. "Observation of a new particle in the search for the Standard Model Higgs boson with the ATLAS detector at the LHC". In: *Physics Letters B* 716.1 (2012), pp. 1 –29. DOI: `10.1016/j.physletb.2012.08.020`.

[2] Class for Physics of the royal Swedish academy of Sciences. "The Nobel Prize in Physics 2013 - Advanced Information". In: *Nobelprize.org* (2013). URL: `http://www.nobelprize.org/nobel_prizes/physics/laureates/2013/advanced.html`.

[3] C. S. Wu et al. "Experimental Test of Parity Conservation in Beta Decay". In: *Phys. Rev.* 105 (4 Feb. 1957), pp. 1413–1415. DOI: `10.1103/PhysRev.105.1413`.

[4] L. Landau. "On the conservation laws for weak interactions". In: *Nuclear Physics* 3.1 (1957), pp. 127 –131. DOI: `10.1016/0029-5582(57)90061-5`.

[5] V. L. Fitch, R. F. Roth, J. S. Russ, and W. Vernon. "Evidence for Constructive Interference Between Coherently Regenerated and CP-Nonconserving Amplitudes". In: *Phys. Rev. Lett.* 15 (2 July 1965), pp. 73–76. DOI: `10.1103/PhysRevLett.15.73`.

[6] J. W. Cronin. "CP Symmetry Violation - The Search For Its Origin". In: *Nobel lecture* (1980). URL: `http://www.nobelprize.org/nobel_prizes/physics/laureates/1980/cronin-lecture.pdf`.

[7] V. L. Fitch. "The Discovery of Charge - Conjugation Parity Asymmetry". In: *Nobel lecture* (1980). URL: `http://www.nobelprize.org/nobel_prizes/physics/laureates/1980/fitch-lecture.pdf`.

[8] J. R. Fry. "CP violation and the standard model". In: *Reports on Progress in Physics* 63.2 (2000), p. 117. URL: `http://stacks.iop.org/0034-4885/63/i=2/a=202`.

[9] G. Lüders. "Proof of the TCP theorem". In: *Annals of Physics* 2.1 (1957), pp. 1 –15. DOI: 10.1016/0003-4916(57)90032-5.

[10] R. G. Sachs. *The Physics of Time Reversal*. University of Chicago Press, 1987.

[11] S. L. Adler. "Axial-Vector Vertex in Spinor Electrodynamics". In: *Phys. Rev.* 177 (5 Jan. 1969), pp. 2426–2438. DOI: 10.1103/PhysRev.177.2426.

[12] J. S. Bell and R. Jackiw. "A PCAC puzzle: $\pi 0 \gamma\gamma$ in the σ-model". In: *Nuovo Cimento A Serie* 60 (Mar. 1969), pp. 47–61. DOI: 10.1007/BF02823296.

[13] C. A. Baker et al. "An Improved experimental limit on the electric dipole moment of the neutron". In: *Phys.Rev.Lett.* 97 (2006). DOI: 10.1103/PhysRevLett.97.131801.

[14] R. J. Crewther, P. D. Vecchia, G. Veneziano, and E. Witten. "Chiral estimate of the electric dipole moment of the neutron in quantum chromodynamics". In: *Physics Letters B* 88.1–2 (1979), pp. 123 –127. DOI: 10.1016/0370-2693(79)90128-X.

[15] R. D. Peccei. "The Strong CP Problem and Axions". In: *Axions*. Ed. by Markus Kuster, Georg Raffelt, and Berta Beltrán. Vol. 741. Lecture Notes in Physics. Springer Berlin Heidelberg, 2008, pp. 3–17. DOI: 10.1007/978-3-540-73518-2_1.

[16] R. D. Peccei and H. R. Quinn. "CP Conservation in the Presence of Pseudoparticles". In: *Phys. Rev. Lett.* 38 (25 June 1977), pp. 1440–1443. DOI: 10.1103/PhysRevLett.38.1440.

[17] J. E. Kim. "Light pseudoscalars, particle physics and cosmology". In: *Physics Reports* 150.1–2 (1987), pp. 1 –177. DOI: 10.1016/0370-1573(87)90017-2.

[18] F. Wilczek. "Problem of Strong P and T Invariance in the Presence of Instantons". In: *Phys. Rev. Lett.* 40 (5 Jan. 1978), pp. 279–282. DOI: 10.1103/PhysRevLett.40.279.

[19] F. Wilczek. "The birth of axions - a citation-classic commentary on problem of strong p and t invariance in the presence of instantons". In: *Citation-classic, CC/ENG TECH APPL SCI* 16.8-9 (1991). URL: http://garfield.library.upenn.edu/classics1991/A1991FE77000001.pdf.

[20] S. Weinberg. "A New Light Boson?" In: *Phys. Rev. Lett.* 40 (4 Jan. 1978), pp. 223–226. DOI: 10.1103/PhysRevLett.40.223.

[21] P. Sikivie. "The Pool-Table Analogy with Axion Physics". In: *Physics Today* 49.12 (1996), p. 22. DOI: 10.1063/1.881573.

[22] M. A. Shifman, A. I. Vainshtein, and V. I. Zakharov. "Can confinement ensure natural CP invariance of strong interactions?" In: *Nuclear Physics B* 166.3 (1980), pp. 493 –506. DOI: 10.1016/0550-3213(80)90209-6.

[23] M. Dine, W. Fischler, and M. Srednicki. "A simple solution to the strong CP problem with a harmless axion". In: *Physics Letters B* 104.3 (1981), pp. 199 –202. DOI: 10.1016/0370-2693(81)90590-6.

[24] P. Svrcek and E. Witten. "Axions in string theory". In: *Journal of High Energy Physics* 2006.06 (2006), p. 051. URL: http://stacks.iop.org/1126-6708/2006/i=06/a=051.

[25] K. Baker et al. "The quest for axions and other new light particles". In: *Annalen der Physik* 525.6 (2013), A93–A99. DOI: 10.1002/andp.201300727.

[26] J. Jäckel and A. Ringwald. "The Low-Energy Frontier of Particle Physics". In: *Annual Review of Nuclear and Particle Science* 60.1 (2010), pp. 405–437. DOI: 10.1146/annurev.nucl.012809.104433.

[27] J. Redondo and A. Ringwald. "Light shining through walls". In: *Contemporary Physics* 52.3 (2011), pp. 211–236. DOI: 10.1080/00107514.2011.563516.

[28] H. Primakoff. "Photo-Production of Neutral Mesons in Nuclear Electric Fields and the Mean Life of the Neutral Meson". In: *Phys. Rev.* 81 (5 Mar. 1951), pp. 899–899. DOI: 10.1103/PhysRev.81.899.

[29] L. B. Okun. "Limits of electrodynamics: paraphotons?" In: *Journal of Experimental and Theoretical Physics (JETP)* 56 (1982), p. 502. URL: http://www.jetp.ac.ru/cgi-bin/dn/e_056_03_0502.pdf.

[30] S. A. Abel et al. "Kinetic mixing of the photon with hidden U(1)s in string phenomenology". In: *Journal of High Energy Physics* 2008.07 (2008), p. 124. URL: http://stacks.iop.org/1126-6708/2008/i=07/a=124.

[31] M. Goodsell, J. Jäckel, J. Redondo, and A. Ringwald. "Naturally light hidden photons in LARGE volume string compactifications". In: *Journal of High Energy Physics* 2009.11 (2009), p. 027. URL: http://stacks.iop.org/1126-6708/2009/i=11/a=027.

[32] J. Jäckel. "A force beyond the Standard Model - Status of the quest for hidden photons". In: *Frascati Physics Series* 56 (2012), pp. 172–192. arXiv:1303.1821 [hep-ph].

[33] B. Döbrich. "Phenomenology of the vacuum in quantum electro-dynamics and beyond". PhD thesis. Friedrich-Schiller-Universität, Jena, 2011. URL: http://www.db-thueringen.de/servlets/DocumentServlet?id=19891.

[34] J. Khoury and A. Weltman. "Chameleon Fields: Awaiting Surprises for Tests of Gravity in Space". In: *Phys. Rev. Lett.* 93 (17 Oct. 2004), p. 171104. DOI: 10.1103/PhysRevLett.93.171104.

[35] A. Ringwald. "Exploring the role of axions and other WISPs in the dark universe". In: *Physics of the Dark Universe* 1.1–2 (2012), pp. 116 –135. DOI: 10.1016/j.dark.2012.10.008.

[36] G. Bertone, D. Hooper, and J. Silk. "Particle dark matter: evidence, candidates and constraints". In: *Physics Reports* 405.5–6 (2005), pp. 279 –390. DOI: 10.1016/j.physrep.2004.08.031.

[37] R. D. Blandford and R. Narayan. "Cosmological Applications of Gravitational Lensing". In: *Annual Review of Astronomy and Astrophysics* 30.1 (1992), pp. 311–358. DOI: 10.1146/annurev.aa.30.090192.001523.

[38] P. Arias et al. "WISPy cold dark matter". In: *Journal of Cosmology and Astroparticle Physics* 2012.06 (2012), p. 013. URL: http://stacks.iop.org/1475-7516/2012/i=06/a=013.

[39] D. De Angelis, A. Roncadelli, and O. Mansutti. "Evidence for a new light spin-zero boson from cosmological gamma-ray propagation?" In: *Phys. Rev. D* 76 (12 Dec. 2007), p. 121301. DOI: 10.1103/PhysRevD.76.121301.

[40] J. Isern, E. García-Berro, S. Torres, and S. Catalán. "Axions and the Cooling of White Dwarf Stars". In: *The Astrophysical Journal Letters* 682.2 (2008), p. L109. URL: http://stacks.iop.org/1538-4357/682/i=2/a=L109.

[41] G. Raffelt. "Particle physics from stars". In: *Annual Review of Nuclear and Particle Science* 49.1 (1999), pp. 163–216. DOI: 10.1146/annurev.nucl.49.1.163.

[42] J. Isern et al. "White dwarfs as physics laboratories: the case of axions". In: *Proc. of 7th Patras Workshop on Axions, WIMPs and WISPs, Mykonos, Greece* (June 2011), pp. 158–162. arXiv:1204.3565 [astro-ph.SR].

[43] G. Raffelt. "Astrophysical Axion Bounds". In: *Lecture Notes in Physics on Axions*. Vol. 741. Springer Berlin Heidelberg, 2008, pp. 51–71. DOI: 10.1007/978-3-540-73518-2_3.

[44] G. Raffelt. *Stars as laboratories for fundamental physics: The astrophysics of neutrinos, axions, and other weakly interacting particles*. University of Chicago Press, 1996. URL: http://wwwth.mpp.mpg.de/members/raffelt/mypapers/199613.pdf.

[45] P. Sikivie. "Experimental Tests of the Invisible Axion". In: *Phys. Rev. Lett.* 51 (16 Oct. 1983), pp. 1415–1417. DOI: 10.1103/PhysRevLett.51.1415.

[46] K. Ehret et al. "New ALPS results on hidden-sector lightweights". In: *Physics Letters B* 689.4–5 (2010), pp. 149 –155. DOI: 10.1016/j.physletb.2010.04.066.

[47] S. Andriamonje et al. "An improved limit on the axion–photon coupling from the CAST experiment". In: *Journal of Cosmology and Astroparticle Physics* 2007.04 (2007), p. 010. URL: http://stacks.iop.org/1475-7516/2007/i=04/a=010.

[48] M. Arik et al. *CAST solar axion search with He^3 buffer gas: Closing the hot dark matter gap*. 2013. arXiv:1307.1985 [hep-ex].

[49] S. J. Asztalos et al. "SQUID-Based Microwave Cavity Search for Dark-Matter Axions". In: *Phys. Rev. Lett.* 104 (4 Jan. 2010), p. 041301. DOI: 10.1103/PhysRevLett.104.041301.

[50] L. Rosenberg and K. Van Bibber. "Searching for Dark Matter Axions". In: *Beam Line* 27.3 (1997). URL: http://www.slac.stanford.edu/pubs/beamline/27/3/27-3-rosenberg.pdf.

[51] P. L. Slocum et al. "Search for a Scalar Axion-like Particle at 34 GHz". In: *Proc. of 8th Patras Workshop on Axions, WIMPs and WISPs, Chicago, IL* (July 2012). arXiv:1301.6184 [astro-ph.CO].

[52] A. T. Malagon et al. "Yale Microwave Cavity Experiment". In: *Proc. of 9th Patras Workshop on Axions, WIMPs and WISPs, Mainz, Germany* (June 2013).

[53] K. Van Bibber et al. "Proposed experiment to produce and detect light pseudoscalars". In: *Phys. Rev. Lett.* 59 (7 Aug. 1987), pp. 759–762. DOI: 10.1103/PhysRevLett.59.759.

[54] M. Gasperini. "Axion production by electromagnetic fields". In: *Phys. Rev. Lett.* 59 (4 July 1987), pp. 396–398. DOI: 10.1103/PhysRevLett.59.396.

[55] F. Hoogeveen and T. Ziegenhagen. "Production and detection of light bosons using optical resonators". In: *Nuclear Physics B* 358.1 (1991), pp. 3 –26. DOI: 10.1016/0550-3213(91)90528-6.

[56] A. S. Chou et al. "Search for Axion like Particles Using a Variable-Baseline Photon-Regeneration Technique". In: *Phys. Rev. Lett.* 100 (8 Feb. 2008), p. 080402. DOI: 10.1103/PhysRevLett.100.080402.

[57] P. Pugnat et al. "Results from the OSQAR photon-regeneration experiment: No light shining through a wall". In: *Phys. Rev. D* 78 (9 Nov. 2008), p. 092003. DOI: 10.1103/PhysRevD.78.092003.

[58] A. Cadène et al. *Vacuum magnetic linear birefringence using pulsed fields: the BMV experiment.* Feb. 2013. arXiv:1302.5389.

[59] R. Bähre et al. "Any light particle search II — Technical Design Report". In: *Journal of Instrumentation* 8.09 (2013), T09001. URL: http://stacks.iop.org/1748-0221/8/i=09/a=T09001.

[60] F. Hoogeveen. "Terrestrial axion production and detection using RF cavities". In: *Physics Letters B* 288.1-2 (1992), pp. 195 –200. DOI: 10.1016/0370-2693(92)91977-H.

[61] J. Jäckel and A. Ringwald. "A cavity experiment to search for hidden sector photons". In: *Physics Letters B* 659.3 (2008), pp. 509 –514. DOI: 10.1016/j.physletb.2007.11.071.

[62] F. Caspers, J. Jäckel, and A. Ringwald. "Feasibility, engineering aspects and physics reach of microwave cavity experiments searching for hidden photons and axions". In: *Journal of Instrumentation* 4.11 (2009), P11013. URL: http://stacks.iop.org/1748-0221/4/i=11/a=P11013.

[63] R. G. Povey, J. G. Hartnett, and M. E. Tobar. "Microwave cavity light shining through a wall optimization and experiment". In: *Phys. Rev. D* 82 (5 Sept. 2010), p. 052003. DOI: 10.1103/PhysRevD.82.052003.

[64] A. Wagner et al. "A Search for Hidden Sector Photons with ADMX". In: *Phys.Rev.Lett.* 105 (2010), p. 171801. DOI: 10.1103/PhysRevLett.105.171801. arXiv:1007.3766 [hep-ex].

[65] M. Kalliokoski et al. "Status of the CASCADE microwave cavity experiment". In: *Proc. of the 9th Patras Workshop on Axions, WIMPs and WISPs, Mainz, Germany* (2013).

[66] M. V. Romalis and R. R. Caldwell. *Laboratory search for a quintessence field.* 2013. arXiv:1302.1579 [astro-ph.CO].

[67] P. Sikivie, D. B. Tanner, and K. van Bibber. "Resonantly Enhanced Axion-Photon Regeneration". In: *Phys. Rev. Lett.* 98 (17 Apr. 2007), p. 172002. DOI: 10.1103/PhysRevLett.98.172002.

[68] P. Sikivie, N. Sullivan, and D. B. Tanner. *Axion Dark Matter Detection using an LC Circuit.* 2013. arXiv:1310.8545 [hep-ph].

[69] F. Wilczek. "Two applications of axion electrodynamics". In: *Phys. Rev. Lett.* 58 (18 May 1987), pp. 1799–1802. DOI: 10.1103/PhysRevLett.58.1799.

[70] F. Marcel. "High-energy physics in a new guise". In: *Physics* 1 (Nov. 2008), p. 36. DOI: 10.1103/Physics.1.36.

[71] L. Rundong, W. Jing, Q. Xiao-Liang, and Z. Shou-Cheng. "Dynamical axion field in topological magnetic insulators". In: *Nat Phys* 6 (4 Apr. 2010), p. 284. DOI: 10.1038/nphys1534.

[72] P. Arias and A. Ringwald. "Illuminating WISPs with Photons". In: *Proc. of PHOTON2011, SPA, Belgium* (May 2011). arXiv:1110.2126 [hep-ph].

[73] B. Aune et al. "Superconducting TESLA cavities". In: *Phys. Rev. ST Accel. Beams* 3 (9 Sept. 2000), p. 092001. DOI: 10.1103/PhysRevSTAB.3.092001.

[74] R. V. Pound. "Electronic Frequency Stabilization of Microwave Oscillators". In: *Review of Scientific Instruments* 17.11 (1946), pp. 490–505. DOI: 10.1063/1.1770414.

[75] T. Mastoridis et al. "Radio frequency noise effects on the CERN Large Hadron Collider beam diffusion". In: *Phys. Rev. ST Accel. Beams* 14 (9 Sept. 2011), p. 092802. DOI: 10.1103/PhysRevSTAB.14.092802.

[76] D. M. Pozar. *Microwave Engineering*. 3rd. Wiley India Pvt. Limited, 2009.

[77] H. Henke. "Basic Concepts I and II". In: *CERN Accelerator School on Radio Frequency Engineering*. 2005. URL: http://cds.cern.ch/record/386544/.

[78] H. L. Thal. "Cylindrical TE011 /TM111 Mode Control by Cavity Shaping". In: *IEEE Transactions on Microwave Theory and Techniques* 27.12 (1979), pp. 982–986. DOI: 10.1109/TMTT.1979.1129777.

[79] H. R. Allan and C. D. Curling. "The design and use of resonant cavity wavemeters for spectrum measurements of pulsed transmitters at wavelengths near 10 cm". In: *Electrical Engineers - Part III: Radio and Communication Engineering, Journal of the Institution of* 95.38 (1948), pp. 473–484. DOI: 10.1049/ji-3-2.1948.0117.

[80] D. Kajfez. "Linear fractional curve fitting for measurement of high Q factors". In: *IEEE Transactions on Microwave Theory and Techniques* 42.7 (1994), pp. 1149–1153. DOI: 10.1109/22.299749.

[81] X. C. Tong. *Advanced Materials and Design for Electromagnetic Interference Shielding*. Taylor & Francis, 2008.

[82] P. Dixon. "Cavity-resonance dampening". In: *IEEE Microwave Magazine* 6.2 (2005), pp. 74–84. DOI: 10.1109/MMW.2005.1491270.

[83] K. H. Gonschorek and H. Singer. *Elektro-Magnetische Verträglichkeit. Grundlagen, Analysen, Maßnahmen*. B.G. Teubner Stuttgart, 1992.

[84] E. Daw and R. F. Bradley. "Effect of high magnetic fields on the noise temperature of a heterostructure field-effect transistor low-noise amplifier". In: *Journal of Applied Physics* 82.4 (1997), pp. 1925–1929. DOI: 10.1063/1.366000.

[85] P. Wade. "Waveguide Filters You Can Build — and Tune. Part 3 Evanescent Mode Waveguide Filters". In: *ARRL Periodicals, QEX Issue March* (Mar. 2010). URL: http://www.w1ghz.org/filter/QEX_Mar_2010_p23-29.pdf.

[86]   Agilent Technologies. *Fundamentals of RF and Microwave Noise Figure Measurements*. Application Note 57-1. 2010. URL: `http://literature.agilent.com/litweb/pdf/5952-8255E.pdf`.

[87]   Agilent Technologies. *Noise Figure Measurement Accuracy – The Y-Factor Method*. Application Note 57-2. 2010. URL: `http://literature.agilent.com/litweb/pdf/5952-3706E.pdf`.

[88]   Agilent Technologies. *Vector Signal Analysis Basics*. Application Note 150-15. 2004. URL: `http://cp.literature.agilent.com/litweb/pdf/5989-1121EN.pdf`.

[89]   A. J. Viterbi. *Principles of coherent communication*. McGraw-Hill series in systems science. McGraw-Hill, 1966.

[90]   G. Heinzel, A. Rüediger, and R. Schilling. *Spectrum and spectral density estimation by the Discrete Fourier transform (DFT), including a comprehensive list of window functions and some new flat-top windows*. MPI für Laser Interferometry & Gravitational Wave Astronomy, internal note. Feb. 2002. URL: `http://edoc.mpg.de/395068`.

[91]   T. S. Durrani and J. M. Nightingale. "Probability distributions for discrete Fourier spectra". In: *Proceedings of the Institution of Electrical Engineers* 120.2 (1973), pp. 299–311. DOI: `10.1049/piee.1973.0062`.

[92]   P. E. Johnson and D. G. Long. "The probability density of spectral estimates based on modified periodogram averages". In: *IEEE Transactions on Signal Processing* 47.5 (1999), pp. 1255–1261. DOI: `10.1109/78.757213`.

[93]   A. Papoulis and S. U. Pillai. *Probability, Random Variables And Stochastic Processes*. McGraw-Hill Series in Electrical Engineering: Communications and signal processing. McGraw Hill, 2002.

[94]   T. H. Joo and A. V. Oppenheim. "Effects of FFT coefficient quantization on sinusoidal signal detection". In: *Acoustics, Speech, and Signal Processing, 1988. ICASSP-88., 1988 International Conference on*. 1988, 1818–1821 vol.3. DOI: `10.1109/ICASSP.1988.196975`.

[95]   Agilent Technologies. *Signal Source Solutions for Coherent and Phase Stable Multi-Channel Systems*. Application Note 59-54. 2010. URL: `http://cp.literature.agilent.com/litweb/pdf/5990-5442EN.pdf`.

[96] P. Daly and I. D. Kitching. "Characterization of NAVSTAR GPS and GLONASS on-board clocks". In: *IEEE Aerospace and Electronic Systems Magazine* 5.7 (1990), pp. 3–9. DOI: 10.1109/62.134214.

[97] M. Frigo and S. G. Johnson. "The Design and Implementation of FFTW3". In: *Proceedings of the IEEE* 93.2 (2005). Special issue on "Program Generation, Optimization, and Platform Adaptation", pp. 216–231.

[98] H. C. Thomas and M. N. David. "Performing out-of-core FFTs on parallel disk systems". In: *Parallel Computing* 24.1 (1998), pp. 5–20. DOI: 10.1016/S0167-8191(97)00114-2.

[99] D. H. Bailey. "FFTs in external or hierarchical memory". In: *Journal of Supercomputing* 4 (1990), pp. 23–35. URL: http://www.davidhbailey.com/dhbpapers/fftq.pdf.

[100] M. R. Portnoff. "An efficient method for transposing large matrices and its application to separable processing of two-dimensional signals". In: *IEEE Transactions on Image Processing* 2.1 (1993), pp. 122–124. DOI: 10.1109/83.210874.

[101] M. Betz and F. Caspers. "A microwave paraphoton and axion detection experiment with 300 dB electromagnetic shielding at 3 GHz". In: *Proc. of International Particle Accelerator Conference* (2012). arXiv:1207.3275 [physics.ins-det].

[102] M. Betz, F. Caspers, M. Gasior, and M. Thumm. "Status report and first results of the microwave LSW experiment at CERN". In: *Proc. of 8th Patras Workshop on Axions, WIMPs and WISPs* (2012). arXiv:1309.7373 [physics.ins-det].

[103] M. Betz, F. Caspers, and M. Gasior. "Status report of the CERN microwave axion experiment". In: *Proc. of 9th Patras Workshop on Axions, WIMPs and WISPs* (2013). arXiv:1309.0341 [physics.ins-det].

[104] M. Betz et al. "First results of the CERN Resonant Weakly Interacting sub-eV Particle Search (CROWS)". In: *Phys. Rev. D* 88 (7 Oct. 2013), p. 075014. DOI: 10.1103/PhysRevD.88.075014.

[105]   P. E. Faugeras et al. "The new RF system for Lepton acceleration in the SPS". In: *Proc. of the 1987 IEEE Particle Accelerator Conference* (1987). URL: http://accelconf.web.cern.ch/accelconf/p87/PDF/PAC1987_1719.PDF.

[106]   G. Prior. *A possible test of MICE RF cavities in a magnet at CERN.* Talk at NEU2012 meeting. 2010. URL: http://indico.cern.ch/getFile.py/access?contribId=27&sessionId=3&resId=0&materialId=slides&confId=106198.

[107]   E. Montesinos. "Private communication". 06/2013.

Curriculum Vitæ

Michael Betz

Michael Betz

Personal Information

Date / place of birth	13th May 1987 / Reutlingen, Germany
Nationality	German

Education

01/2011 - 01/2014 — **PhD**, *Electrical Engineering*, Karlsruhe Institute of Technology / CERN.

Scholarship — Supported by the "Wolfgang Gentner Scholarship" of the German Federal Ministry of Education and Research (BMBF)

03/2009 - 10/2010 — **Master of Engineering**, *Electrical Engineering*, Hochschule Karlsruhe.
Emphasis on: *Information and communication technology*

Thesis — *Feasibility Study for High Power RF – Energy Recovery in Particle Accelerators*, at CERN, Geneva.

Key Work — The feasibility of solid state rectifier modules to recover RF power in the MW range for certain particle accelerators is discussed.

10/2005 - 2/2009 — **Bachelor of Engineering**, *Mechatronik*, Hochschule Reutlingen.
Emphasis on: *Automation*

09/2003 - 07/2005 — **Matriculation standard**, *Ferdinand von Steinbeis Schule*, Reutlingen.
Emphasis on: *Information and communication technology*

Invited talks, research visits and further activities

12/2013 — **Invited talk**, *Results of the CERN Resonant WISP Search (CROWS)*, DESY Physics Seminar, Hamburg, Zeuthen.

06/2013 — **Invited talk**, *Cavity WISP searches at CERN*, Dark Matter: A Light Move, DESY, Hamburg.

10/2012 — **Invited talk**, *Axion-like particle and hidden photon search with a microwave shining through the wall experiment*, ALPS - Seminar, DESY, Hamburg.

07/2012 — **Invited research visit (10 days)**, *Yale Microwave Cavity Experiment (YMCE)*, Collaborative exchange, New Haven (CI).

Since 02/2011 — **Joint Universities Accelerator School**, Archamps, France.
Assisting the lectures as well as the theoretical and practical exercises on Radio-Frequency engineering

Since 07/2011 — **Electron cloud measurements**, CERN, Geneva.
Assistance in prototyping a microwave diagnostic system in the Super Proton Synchrotron for real-time measurement of electron cloud formation by the multipacting effect

KARLSRUHER FORSCHUNGSBERICHTE AUS DEM
INSTITUT FÜR HOCHLEISTUNGSIMPULS- UND MIKROWELLENTECHNIK
(ISSN 2192-2764)